CW01508627

THE TRAVELLERS' LIBRARY
★
A GIPSY OF THE HORN

¶ A descriptive list of some of the volumes in
  THE TRAVELLERS' LIBRARY will be
  found at the end of the volume.
¶ As further volumes are constantly being added
  to the Library it is not always possible to keep
  these lists fully up to date.   For the latest lists
  application should be made to any bookseller,
  or to the publishers.

A GIPSY OF THE HORN

The narrative of a voyage round the
world in a Windjammer twenty years ago

by

REX CLEMENTS

LONDON
JONATHAN CAPE 30 BEDFORD SQUARE

FIRST PUBLISHED 1924
FIRST ISSUED IN THE TRAVELLERS' LIBRARY 1929

PRINTED IN GUERNSEY, C.I., BRITISH ISLES

FOREWORD

*

Through the courtesy of Messrs. Heath Cranton Limited, by whom this book was published in 1924, I have been enabled to prepare this new and revised edition for the " Travellers' Library."

I trust that old readers of the book will welcome it in its shorter form, and that new readers may enjoy 'a plain tale of the sea.'

REX CLEMENTS

CONTENTS

A

A GIPSY OF THE HORN

CHAPTER ONE
Outward Bound
*

THE romance of a sea-life has ever been a potent factor in the making of sailors. It was so in my case: romance and much reading of *Robinson Crusoe* were my god-fathers in the matter. The strange surprising adventures of Defoe's immortal mariner fired my imagination, and so irresistible did the beauty of white-winged ships and the wonder of far lands become, that at an age when most boys are supposed to have attained years of discretion, nothing would satisfy me but the life of a sailor.

Fate at length gave way to my importunities, and a discussion of ways and means followed. After much deliberation and close scrutiny of a list sent us by a marine agent, containing the names of eleven ships whose owners were all willing to take apprentices, a small Scottish barque called the *Arethusa* was selected. The name, I think, was the deciding factor; it smacked of the sea and suggested a saucy frigate and salt adventure generally. Besides, the god of ocean had befriended the nymph Arethusa of old, and might he not prove as propitious to her modern namesake? Anyhow, the *Arethusa* it was, and afterwards I had reason to be thankful for the choice, for at the end of my apprenticeship she was the only one of those eleven ships that remained afloat – all the others had been lost.

I soon made the acquaintance of the vessel that was to carry me beyond the skyline. Indentured and brass-

bound, I joined her one wintry January afternoon as she
lay in the East India Docks, London — fully loaded and
ready to sail on the morrow for Australian ports. My
first impression, as I turned the corner of a warehouse
and came in sight of her, was one of complete surprise.
Different indeed was the stark reality from the richly-
coloured pictures I had painted. No dashing frigate
or golden galleon was this, that lay with her gaunt spars
towering up into the grey sky above the cranes and
dock-warehouses, her decks littered with the accumu-
lated rubbish of a long stay in port, and the grime of
London over all.

As I clambered aboard and stood looking round, I
was hailed by a man in a blue pilot-jacket, who inquired
my business. On learning I was a new apprentice —
indeed, I looked it — he informed me he was the Third
Mate and the only officer on board. He took me along
to the half-deck, where, he told me, the apprentices
berthed.

I found my new home was a bare, box-like apartment
about ten feet square, with iron walls and wooden bunks
round three of the sides. It was half-filled with coals,
firewood, odds and ends of rope and miscellaneous
rubbish, and looked inexpressibly cheerless.

From the half-deck the Third Mate took me all
round the ship. Everything seemed strange, serviceable
and enormously strong, but bleak and bare as cold steel
and sinewy wire could make it. The most fascinating
thing was an Oriental-like aroma that seemed to per-
meate every corner of the ship, but when I mentioned it

to the Third Mate he laughed: 'That's the guano,' he said, 'she's just come home with a load of it from the Chinchas.'

Fore and aft we went – from the foc'sle head, where I peered down at the sharp cutwater, to the poop, where I fingered the five-foot wheel. My guide was obviously fond of the ship, and more than once pointed with pride to some wide stretch of spar or shapely curve of waterline. The rubbish that littered every corner he dismissed with a 'You'll be able to eat your dinner off these decks before we've been at sea a week,' but in spite of his hearty bearing and evident sincerity I felt a little chilled and disappointed at it all.

Nor was my first meeting with the Captain a few minutes later at all reassuring. The latter called me aft as he came aboard and asked a few questions, winding up by inquiring whether I wanted to go to sea. I replied that I did, whereupon he regarded me sternly and told me I should 'do better to buy a rope and hang myself.' He was a very big, broad-shouldered, weather-beaten man, with a tremendous voice and an impressively large manner, and the interview left me rather crushed.

I found my way back to the half-deck, and without enthusiasm began to clear up some of the rubbish that littered it. In a few minutes I was joined by another boy, who told me his name was Gilroy, and that he also was an apprentice. A third arrival came soon after, thick-set and freckled, who announced himself as Jimmy Rollins and another first voyager. Together we set to work to get the place into some sort of order and make room to unpack our bags and sea-chests.

We were busy at it when a thunderous tattoo was sounded on the iron bulkhead that separated us from the galley, and, running out to see what was the matter, found it was the steward's method of informing us tea was ready. The steward was a tight, compactly-built little man, very hairy, very cheerful and immoderately energetic, clad — in spite of the season — in only a vest and trousers, with his bare feet thrust into a pair of canvas slippers. He was thumping the bulkhead lustily with a saucepan. 'Tea-oh!' he called when he saw us, and passed out a large tin of tea and a similar one containing a quantity of greasy green stew. On closer examination of the fare provided we sipped some of the liquid, but forbore to venture further and made our meal off a few biscuits and cakes we had brought with us.

All the evening we spent unpacking and making, as we imagined, our cabin snug and comfortable for a sea-voyage. We were joined after tea by another brass-bound newcomer, Beckett by name, and with his arrival our muster in the half-deck was complete.

The Third Mate, Mr. Patrick, looked in later and gave us a few hints. Under his guidance we sallied out into the East India Dock Road and made some purchases which he said would be useful, soap and matches among other things, and a lamp for the bulkhead. We finished up with a good supper at a small restaurant near the docks. Though we didn't know it, it was the last square meal we were going to have for a long while, and what with the fare and the Third's enlivening conversation, we were a cheery party by the time we returned to the ship. Back on board, we laid out our working-clothes in readiness for the morrow, and, turning into our strange box-like bunks, fell asleep.

Very early next morning we were aroused by a stentorian 'Way-ay-ay! turn oot ther', you sleepers!' and, sitting up with something of a start, saw a rugged face and brawny pair of shoulders framed in the doorway. They belonged to the ship's carpenter, who came in and turned up the lamp for us. We jumped briskly out of our bunks and began to don our new, stiffly-uncomfortable clothes. The brief enthusiasm of overnight evaporated in the biting air of morning, and Beckett muttered something about 'not caring for a sea-life' as he crawled out of his bunk. Even at the moment we thought the opinion premature, but as far as the speaker was concerned it was prophetic too, sure enough.

Getting in one another's way, stumbling against unaccustomed angles, and complaining loudly at misfits in boots and clothing, we yet managed to get dressed in the regulation half-hour, and a mug of steaming coffee, brought in by the steward, warmed and woke us up. Then – 'Clang-clang! clang-clang!' – four bells was struck somewhere, and we stumbled out on deck.

It was still quite dark, and as I became aware of my surroundings my only feeling was one of utter bewilderment. Overhead the great masts and yards and spidery web of the rigging loomed dim and unreal in the flickering light of a few gas-lamps on the quay. A tangle of

ropes and wires littered the deck, and a diminutive tug
was puffing and blowing noisily alongside. In the fore
part of the ship, some of the crew, who had come on
board during the night, were slowly hauling in a clank-
ing mooring-chain. We tallied on behind them and
hauled lustily, and, after that, at another, and another,
and another. It was cold, wet work; the men seemed
half-asleep, and Mr. Patrick and a grizzled old veteran,
who, I was told, was the First Mate, were in the shortest
of tempers.

For a couple of hours, sometimes at the Mate's curt
bidding, sometimes on our own initiative, we ran hither
and thither, hauling here and hauling there, not understanding what it was all about, but perceiving that slowly
the ship moved away from the quay and out into the
dock.

It grew light, and still we pulled and hauled, handling
wet and icy ropes and endless greasy wires possessed by
a very demon of unexpected spitefulness. Our backs
ached, our hands blistered, and we grew more and more
caked with rich Thames mud.

After what seemed an age of weary and bewildering
labour we found that the *Arethusa* was passing steadily
into the narrow lock that afforded access to the river.
It was broad daylight by this time and, as soon as the
ship was snugly in, order was given to make fast, and all
hands were sent to breakfast.

Visions of a white cloth, of hot rolls and buttered
scones, that the welcome word conjured up, soon faded
into the naked reality – another tin of tea and that

abominable green stew. Hungry as we were, we left it, and ate a few more of the dainties packed up by sympathetic hands at home.

There followed another spell of hilly-hauling and yo-hoing. The captain appeared on the poop with an elderly, black-coated man; our pilot climbed aboard, and there was a general shouting of orders. The outer lock-gates swung open, and a tug came up and was made fast for'ard. The elderly man, who somebody said was the owner, scrambled ashore; our ensign was dipped; a straggling cheer came from a few onlookers on the pier-head; and, to a vociferous chorus of toots from the tug, the *Arethusa* gathered way and glided out on to the broad bosom of the Thames. We were off !

It was a cold, clear morning, with a fresh breeze ruffling the surface of the river. As the ship's head turned downstream and speed increased, we boys were set to work coiling up ropes and tidying the decks. I collected a big pile of rubbish by the side of the fore-hatch, to be thrown overboard as soon as we were outside the limits of the Thames Conservancy Board.

In the intervals of my labours I watched the changing panorama on either side with interest.

Down Blackwall Reach, famous in song and story, we went; past the Gallions; making numberless bends and turns in our course, while the buildings on either side thinned out and the smoke of London banked up behind.

At Gravesend the tug slowed down, and we changed pilots. Then on again, past the great docks at Tilbury, while the stream broadened out and the character of the river gradually changed.

On either hand the banks receded, and we were now stemming through a turbid, swiftly-moving flood; dull yellow gleams of sand with grey little wavelets lapping over them marked the presence of shoal and sandbank, and the freshening breeze brought with it a tang of salt from the open sea ahead. It was growing dusk as we passed the Nore lightship, and an eerie wind began to whistle through the cordage out of the deepening gloom around.

I was reflecting that it was a grey and mournful sort of night to be setting out on one's adventures, and not at all touched with the picturesqueness of departure, when I was startled by a sudden shout from the poop just above my head: 'All – hands – lay aft!'

A shuffling and trampling of feet from for'ard answered the call. The men gathered in a little knot by the after-hatch, and the First and Second Mates came down the poop ladder. The former had a paper in his hand, and, glancing at it, called a number of names, the responses to which were given in a surprising variety of tongues and accents. Then the two officers commenced to pick their watches, and I had an opportunity to take stock of the men among whom I was to live for the next year or so.

A motley crowd they were, of all shapes, sizes, colours and nationalities, yet somehow they looked different from labouring men ashore. In each and all there was a subtle something that differentiated them from those whose lives were cast in a less spacious environment. Whether it was in their bearing, their easy posture, the forward swing of their arms, their big hands half-clenched through much handling of ropes, or whether it was something deeper that looked out through their eyes, I don't know, but a difference there certainly was. Sailing ships have a way of stamping the men that sail them. The weather-glancing eye, the rolling walk that climbing ratlines gives, the leisurely manner that masks an instant readiness, are common to all who have handled canvas.

As a matter of fact, I was fortunate in the crew I sailed with on this my first voyage. They were distinctly above the average. Two of them would have been an asset in any foc'sle. Of these, one was a typical British sailor of the finest type, Stedman by name, a blue-eyed, broad-shouldered man of about thirty, a sailor every inch of him, able to hand, reef or steer, and master of all an A.B. should know, from sailing a boat to working cunning knots and splices dear to the old-time sailor's heart. The other was a huge, yellow-bearded seaman, a perfect Viking of a man, hairy-chested, mighty-shouldered, with great corded arms, and tattooed all over. His name was MacDonnel – big Mac, we always called him. He was as Irish as the Blarney-stone, and one of the best-tempered and most light-hearted men I ever met. Nothing could upset him, and nothing could daunt him. He was not so all-round a seaman as Stedman, but his strength was prodigious. He could bend an iron belaying-pin between his bare hands, and muscles of that calibre always command respect – anyway on board ship.

In addition to these two there was another Britisher among the crowd – Jamieson, a craggy-featured old Scotch shell-back. He admirably fulfilled the definition of such a one as having 'every finger a fish hook and every hair a rope yarn,' but he was an old man and long past his prime.

The rest were the usual all-nation crowd and hailed from as many countries as there were men. There was Brice, a lean and leathery New Englander, hard as nails and a good man in a tight corner; Nils, a silent Russian Finn, with a figure like a stone bollard; and Neilsen, a strapping pink and white young Norwegian, as active as a cat aloft and with a complexion any society lady might have envied. These were all good seamen and above the average.

Not so good were Lopez – a poor old Spaniard – very willing but very inefficient; Schmidt, a red-nosed old German, and Johnson, a darky, lively enough in fair weather but a crawling bag of jelly in bad.

However, the crowd taken as a whole was distinctly above the average. We had five good sailormen out of eight all told. Compared with some crowds I have sailed with they were all one could wish for, and taken 'by and large' made for a happy ship as only a capable crew can.

The two mates soon picked their watches – Stedman, Mac, Neilsen and Lopez in the Chief's; Jamieson, Brice, Nils, Schmidt and Johnson in the Second's. The muster completed, the mate turned to Beckett and me with a 'come over here, you two! and you' – to Gilroy and Jimmy – 'go into Mr. Miller's watch!' Having thus quickly disposed of us boys the mate turned away saying, 'Go below the port watch!' and, this being his own, we were released for a couple of hours.

By the time we came on deck again it was pitch dark and we seemed to be well out of the river, for nothing was visible except a few lights flashing and dancing in various directions. The tug was still puffing away ahead and, the wind being fair, a sail or two had been set to help us along. The breeze had freshened and was piping and whistling through the rigging, while the air had turned bitterly cold. Finding no one took any notice of us, Beckett and I made our way to the break of the poop, where there was a little shelter from the cutting wind, and talked quietly together, feeling 'at sea' in every sense of the word. We were soon joined by the third mate, who lit his pipe and paced steadily up and down athwartships, stopping to point out to us the light on Cape Gris Nez and the loom of the Foreland flash, and yarning away the while about ships and the sea in a very genial manner.

Mr. Patrick, or 'Paddy' as we more often called him, told us one of our duties in our watch on deck at nights was to strike the bells. A little clock, by which to note the time, hung in the companion, and every half-hour we had to strike the bell on the wheel-box behind the steersman. On hearing it the lookout man replied with a similar number of strokes on the big bell for'ard and

23

the cry of 'All's well, sir.' All time on board ship is reckoned by means of bells struck in this way. 'Struck,' never 'rung', as the Third was careful to point out. There is no such thing as 'ringing' a bell on board ship, except when at anchor in a fog. From one to eight strokes are made, the full number marking the end of the watch and occurring at noon, midnight and four and eight a.m. and p.m.

The dog-watches have a system of their own. From four to six p.m. they follow the usual course, but at half past six, instead of five bells, only one is struck, two at 7 p.m., three at 7.30 and the full eight bells at eight o'clock. The reason for this variation from time-honoured sea-custom is that five bells in the second dog-watch was the concerted signal for the Mutiny at the Nore to begin, and since that historic occasion it has never been struck on board a British ship. A luckless greenhorn may sometimes strike five bells at the for-bidden time and only be made aware he has committed a serious offence by being well-cursed for a 'useless soger.' On American vessels, of course, five, six, and seven bells continue to be struck in the old way.

Once or twice in the course of the watch, on an order from the poop, we helped to haul taut a rope or flatten a sheet, and at eight bells went below for our first four hours sleep.

Hardly, it seemed, had our weary heads touched the pillow than we were roused again and told it was mid-night and our watch on deck. We were still sleepily fumbling with our clothes when there was a shout outside: 'All hands on deck to cast off the tug!' We stumbled out, the change from the stuffy atmosphere of the half-deck almost taking our breath away.

A stinging nor'easterly wind that drew a deep resonant hum from tautened backstay and halliard was sweeping across the white-tipped seas and cut our faces like a whiplash. In an instant all trace of drowsiness had gone. Gasping and clutching for support we struggled up on to the foc'sle head, where a little group of men was clustered. The second mate was in the act of throwing the end of the wire towing-hawser off the fore bitts. There was a shout of 'Stand clear all!' and with a spiteful 'zip' the end of the hawser flashed round the bitts, raced through the fair-lead and fell into the sea beneath the bows.

'All gone the tow-rope,' yelled the second mate, making a funnel of his hands and shouting into the darkness ahead, where a tiny dancing light marked the presence of the tug.

'Right-O!' was borne faintly back to us, 'pleasant voyage to you!' The dancing light sheered away from the bows: the men turned to leave the foc'sle head, and there came a sharp order from aft: 'Loose the fore and main t'gans'ls'!'

At the word a few men sprang into the weather rigging and swarmed nimbly up. Glancing aloft, I saw that a few strips of sail were already set and could distinguish, high above them, dark figures moving out on the dimly-swaying yards. A moment later the sails fell in heavy folds, that flapped and bellowed furiously in the howling

25

wind. Another sharp order from the poop and the men remaining on deck began to haul out the sheets and spread the thunderous canvas to the wind. They worked with a weird, wailing cry and I hurried to the end of the rope and hauled lustily, glad to do something to warm my numbed limbs. We gave a succession of sharp, violent pulls, the rope coming in in short jerks.

'Belay that!' said the mate, suddenly. 'Halyards!' The sheet was made fast to a belaying-pin and we moved across to the halyards on the other side of the deck.

'Hoist away!' shouted the mate.

One of the men – Mac, I think it was – in a mournful wavering voice began to sing the words of a song, of which the refrain at the end of each line was 'Blow, boys, blow,' chanted by all the men in unison, who hauled as they sang. It had an inexpressibly wild and haunting effect and was my first acquaintance with a real 'shanty' – the famous folk-songs of the deep sea. The last few pulls were given to a succession of short, quick cries –

'Oh, bust her!'

'Strike a light!'

'Now, bullies, when she's not lookin'!'

'Two blocks, sir!' – and the yard being sufficiently hoisted, the order was given to belay.

Following the t'gans'l's, other sails were set in quick succession, until a towering, booming press of canvas was piled above our heads. The ship lay over at a smart angle and a sharp unending hiss rose from the rushing water overside, punctuated by the quick clatter of falling sprays tumbling over the weather bulwarks. It must have been fully half past one when the mate gave the order 'Go below the watch,' and the starbowlines trooped off to get what rest they could before four o'clock.

It was the beginning of a splendid run down Channel. We had found a lucky slant of wind and full advantage was taken of it 'to carry on, and thrash her out with all she'll stand.'

When day dawned, the land lay like a misty blue line on our starboard hand, the fresh breeze was whitening the tops of the steel-grey Channel waves, and the ship, leaning sharply under a towering pyramid of canvas, was almost smoking through the water. It was good to be alive on such a morning, for here was one's excited imagination of a sea-life come true. A smother of foam beneath our bows, a rainbow of spray a-sparkle in the fore-rigging, and the sonorous hum of the wind aloft — what more could the heart of adventure desire? Everybody was delighted at getting clear of the dreaded Channel — the Sea of Sore Heads and Sore Hearts — in such fine style, and I saw the old man cast many approving glances around as he paced steadily up and down the poop.

In the forenoon watch we passed a large four-masted sailing-ship. She was going the same way as ourselves and grandly shouldering the seas aside. Word went round it was the *Lynton*, which had left London on Christmas Eve — three weeks before we did — and had been lying wind-bound in the Downs ever since. The

Lynton it proved to be, and we slashed past, half a mile distant, doing three knots to her two. She shook her royals out as we drew ahead, but to no purpose. One of our men derisively waved the end of a line as a tow-rope, and, though the distance was too great for him to be heard, yelled an offer to report them when we reached Adelaide.

All day the old man paced the weather side of the poop, only going below for a few minutes at a time. On one occasion, before he went down, he turned round at the head of the companion-way and called out in a tremendous voice: 'Boy!'

'Sir,' said I, scrambling up the poop ladder and running to where he was standing.

'Fetch me a hammer from the carpenter,' said he. 'Aye, aye, sir,' and I ran for'ard, obtained the hammer and scrambled aft again as quickly as I could, for the ship's motion was very lively and I had not yet got my sea-legs. The old man was still standing by the companion – very huge, grim and imposing. 'The hammer, sir,' said I mildly.

He took it and regarded me with a stern gaze. 'When I tell you to do a thing, my boy,' he said – and his voice, slow and deep-chested at first, rose higher with each word in a detonating thunder-roll of sound – 'you won't run, you'll —— well JUMP!'

I jumped then, and stared blankly after his broad back as it retreated down the companion. Good heavens! thought I, what a commander to sail under! I never guessed sea-captains were like this.

It was my first encounter with the captain and filled me with misgivings. I was soon to find my first impressions of him were unduly pessimistic, and it did not take long for me to become accustomed to his habit of giving orders in a voice like the flap of a sail, with the last word a regular squall-burst in its vehemence. I grew to admire and like him heartily, though it was long before I was in a position to appreciate his skill and mastery of seamanship. This took time – his breezy truculence was obvious from the outset. The very next day provided another instance of it.

We were washing down the poop, and Stedman, who was handling the buckets, happened to throw some water over Gilroy's legs and begged the latter's pardon. Such consideration annoyed the old man. 'Why haven't you got your oilskins on, boy?' he demanded, 'damme! you want to get wet, do you?' and thereupon he picked up a brimming bucket of water and flung the contents all over the surprised Gilroy. It being a rigid January evening, with a biting nor'easter blowing, it was a chilly baptism for that unfortunate youth. He was blue and chattering before the job was finished and he could get along to the half-deck and change. We gloomily compared notes, he and I, and agreed the old man was a terror.

All this time we were thrashing merrily down Channel and having a grand run of it. It came up to expectation and a bit over. It was wild; it was free; every moment had its own zest, but it was not altogether a dream of joy. For one thing, the food – the little there was of it — was horrible; we mostly dumped it overboard straight away and subsisted on the small stock of luxuries we had brought away with us.

Another unpleasantness was the surprising number of cuts and bruises we managed to pick up. Our hands were raw and blistered through the unaccustomed handling of rough, wet ropes; and the surprising rolls and plunges of the ship were constantly bringing us into painful communication with every point and angle about the decks.

But it was not these things that worried us ; it was the longing to sleep. Four hours on and four hours off sounds easy enough, but to a boy fresh from home and aching from the hard work it was agony. At any moment I could have laid myself down on the bare decks and gone to sleep with the water splashing over me.

And what made it worse was a ridiculous determination on my part to be a real sailor. I had read in some alleged sea-story that no 'sailor' ever slept in his watch below in the day-time. No! the author implied he carved models or knitted jerseys or rioted in other fantastic pastimes. So I — who was going to be a good sailor — sat miserably on my sea-chest, with my legs dangling in the water that washed to and fro across the floor, and fought an agonized fight against sleep. How long it would have lasted I don't know, but on the second or third day the Third Mate looked in and saw me dejectedly keeping my vigil.

'Hello!' he said, 'what the devil are you doing?'

I told him I was trying to keep awake.

'Holy smoke!' said he, 'what for? Got a bet on?'

'No,' said I, 'but I'm sailor enough not to sleep in the day-time.'

The Third laughed: 'Who's been pulling your leg?' said he; 'turn in, you flaming idiot, eight bells 'll be here soon enough.'

He banged the door to and went off laughing. I realized there was something wrong with my information and thankfully rolled into my bunk, fully-dressed as I was, hardly stopping to anathematize the author of that pernicious book.

We passed a large steamship in the course of the night – we had seen surprisingly few in our run down Channel. She was a very big vessel and swept past close by, her hull lit up by long tiers of blazing lights. The Third said she was probably a Union-Castle liner.

When day broke, broad out on our starboard beam lay a long stretch of pleasant coast running out into a bold rugged headland crowned by a white lighthouse. It was the Lizard. We had run 280 miles in the previous twenty-four hours. It was a rattling good start and all hands were in high spirits. The Narrow Seas now lay astern, while under our plunging forefoot stretched the long grey leagues of the Atlantic.

All that day we were employed in catting and fishing the anchors, that is, hoisting them inboard and lashing them down securely on the deck of the foc'sle head. The operation marked our final severance from the land and spelt good-bye to ports and harbours and hey! for the open sea. An arduous undertaking it was and not without an element of danger. The anchors weighed a couple of tons apiece and had first to be hoisted to the cathead – a strong baulk of timber projecting from the side of the bows – and then, by means of a powerful tackle from the foretopmast head, lifted bodily aboard, lowered on deck and firmly lashed in their places. It was a job that called for coolness and skill, and we boys only gasped through the flying sheets of spray and hauled or heaved at the capstan as we were told.

Under the mate's direction Stedman went over the side in a bowline and hooked on the catfall. Slowly, under absolute control, first one anchor and then the other was slung inboard and lashed with chains to iron ring-bolts in the deck.

It was a wet and windy struggle, exhilarating too, and once Gilroy's excitement got the better of him. Suddenly, under our bows, the sea broke white with the rush of a tremendous number of big fish, with triangular fins and curved black gleaming bodies.

'Look!' shouted Gilroy, 'sharks!'

It would have brought him a sharp reprimand a couple of days later. To speak on duty, except in answer to an order, was a breach of discipline and asking for trouble. At the moment neither of us was aware of the strictness of the discipline under which we were to serve and Gilroy's ignorance saved him. As it was the mate only said, 'They're not sharks, they're porpoises.'

Land faded out of sight in the course of the day, but with the deepening shades of evening a solitary gleam of light leapt above the skyline astern and flashed at intervals across the cloudy heavens. It was the flash light on the Bishop Rock in the Scillies and our last sight of old England; we watched it for long, but before four bells struck and it was time to go below, it had finally sunk from sight. The ring of the horizon was unbroken; we were fairly out on the wide sea, and our long voyage had begun — .

> 'the last, last flicker goes
> From the tumbling water-rows,
> And we're off to Mother Carey,
> (Walk her down to Mother Carey),
> Oh, we're bound for Mother Carey where she feeds
> her chicks at sea.'

CHAPTER TWO
Across the Bay
*

As if to assure us that we had got into his own domain at last, the North Atlantic determined to introduce himself. All day the wind had been rising; the light sails had been taken in as soon as the anchors were secured, and at four bells in the first dog-watch, in a ringing hail-squall, the t'gallants'ls were clewed up. Still, with steady persistence, the wind rose. Little seas slopped over the weather-rail and scurried viciously to leeward, and the lurches of the barque became ever more sudden and violent.

In the half-deck there was nothing but sodden, cheerless confusion. The floor was awash and the remains of our tea littered a sea-chest. Beckett had taken to his bunk and was groaning dismally in the throes of seasickness. The Second Mate had told Gilroy to 'get to hell out of it' and he and I were sitting on the edges of our bunks, talking fitfully to one another and listening to the rising shriek of the wind and the savage surge and rush of the sea. It was a dreary evening and our mood was in keeping with it.

As two bells struck, the craggy features of our old Scotch carpenter appeared in the doorway and we hailed him with a hearty 'Come in, Chips,' only too glad to have somebody of experience to talk to. He came in and, sitting down, pulled out his pipe and glanced at our faces, white in the smoky glare of the slush-lamp.

'Ye'll be enjoyin' yersells the noo, ah'm thinkin',' he said drily.

Chips had just finished his tea – or 'supper' as it is always called at sea – and was in a mood for talk. He and the sailmaker, we found, stood no watches but worked all day and had all night in, being known, in consequence, as 'the idlers.' We took advantage of Chips's communicativeness and plied him with all sorts of questions about the ship, the weather and the passage. He replied with deliberation and it was evident that the enjoyment of his after-supper smoke was in no wise impaired by the increasing uproar of the elements without. He was a grim and granite-visaged old Caledonian, with a demeanour as rocky and imperturbable as one of his own Hebrides. But under his gruff exterior there lurked a deal of kindliness. He gave us a number of wrinkles that proved useful, and helped us lash our sea-chests, that were sliding back and fore across the deck, to the imminent hazard of our legs.

'Ah doubt ye'll be takin' the mains'l aff her,' he said, as eight bells was suddenly struck aft and he rose to go. Sure enough, as the big bell for'ard clanged out in answer, came the loud order:

'All hands on deck! Weather main clew garnet!'

We scrambled out on deck and stared into the wet darkness that shut us in on every side. It was a wild-looking prospect – what little we could see of it. High above our bulwarks solid black walls of water, streaked and crested with lines of foam, drove furiously past. Overhead a low sky, packed with racing scud, seemed

35

impendent above our mast-heads. The decks were a
foot deep in swirling water, and in the black caverns
aloft the wind drummed and screamed like a legion of
lost spirits. The hands, in answer to the call, were
stumbling their way aft, their yellow, oilskin-clad
figures gleaming in the wet. Just as I left the shelter of
the house a sousing dollop of water came in over the
weather-rail, swept me off my feet and landed me in the
lee scuppers.

'Why, bully-boy,' said a surprised voice, 'phwat are
yez doin' down there?' and I felt myself grabbed by the
coat and swung to my feet. Looking up, I saw it was
'big Mac.' He was barefooted and without oilskins; a
stringy old blue jersey, with the sleeves cut off at the
elbows, had been drawn on above his one and only shirt,
but his big red face was as smiling as ever.

I ploughed on in Mac's wake to the main rigging
where, under the mates' direction, the men were clearing the ropes and preparing to start the tack.

How that sail was taken in I never realized. It was
like a fight with a living monster. The Second Mate,
shouting orders and cursing vociferously, was slacking
away a rope in the darkness to wind'ard, while the men
manned the clew garnet and buntlines that were to
snug the sail up to the yard. The wet and heavy canvas
banged and bellowed uproariously above our heads as
we hauled away at the lines. Icy sheets of stinging spray
and the tops of waves kept leaping over the weather
bulwarks, sweeping down on us and drenching us to the
skin. Pretty well bewildered as I was, hauling at

seemingly endless ropes, I yet realized that slowly, inch by inch, the sail was being mastered.

Through it all, in language that leapt and flared like a blue flame, hurtled the Second Mate's hurricane orders. I trembled to listen to him. It was the first time I had heard a deep-sea officer with a 'clipper' training take in sail under the eye of his commander, and the language he employed was a revelation to me. On the edge of destruction, as I imagined us to be, here was a man so hardened and careless of his welfare as to be invoking all the deities and devils that ear had ever heard of, with epithets that Ajax defying the lightning would scarcely have used.

At last the tearing, slatting canvas was dragged close up to the yard and the order given:

''Way aloft and furl it!'

I swung myself desperately into the rigging with the idea of helping somehow, but Paddy noticed me:

'Get down, you!' he yelled, 'and coil up those ropes.'

I did as I was told, while the men clambered aloft, fisted the sail and rolled it up on the yard. At the bunt was Mr. Miller and still, above the roar of the wind and the din and clatter of the swinging ports, I could catch fragments of his voice encouraging the men in bursts of frenzied profanity.

As soon as the sail was stowed and the last man down, the order was given to relieve the wheel and look-out, while we of the port-watch were sent aft to spread the weather-dodger. All through the watch we were kept at work attending to one thing or another, and seven

bells had struck before I took up my post at the break
of the poop and was joined, a minute or two later, by
the Third.

The latter sat down on the ledge of the lamp-room
door, pulled off his sea-boots and emptied the water out
of them.

'Well,' said he, 'what d'ye think of a sea-life now?'

'Not too much,' I replied truthfully, 'if it's all like
this.'

Paddy chuckled. 'Why,' said he, 'if we never had a
bit of a blow the girls would be doing us out of our jobs.'

'But surely you'd call this a tempest?'

He shook his head: 'There's no such thing as a
tempest,' said he, 'except in hymns and pray'r books.'

'Well, what would you call it?' I asked, looking at
the great foam-crested seas and the dim outline of the
reeling barque.

'The hell of a dirty night,' said he, and lit his pipe.

We chatted together and I think I must have alluded
somehow to the awful language used by the Second
Mate, for Paddy laughed.

'Oh, that's what's the matter, is it?' said he. 'Did
ever you hear of the parson who made a trip in a lime-
juicer, Clements?'

'No, sir.'

'Well, he did; a proper sky-pilot he was, with a claw-
hammer coat and gafftops'l hat with backstays to it.
He went out in the old *Warrego* and in the Bay they
came in for a bit of a dusting – the same as we're doing.
They shortened sail, and, clewing up, the watch damned

everything from the bilges to breakfast time. It grieved the parson to hear 'em and he went and spoke to the skipper.

'That's all right, sir,' said the old man, 'that's all right. When you hear the men cursing like that there's no fear of the ship going down.'

The parson went below, but popped up again a little later. Pretty bad he was, and praying he'd never come. It was blowing hard – a big sea out of the sou'-west and the ship taking it green. The crowd were aloft, handing the mains'l, and the things they were laying tongue to – well, believe me, you haven't heard swearing yet. That sky-pilot he just listened, took one look round and headed for the companion: 'Thank God, they're cursing still,' says he, 'oh, thank God they're cursing still!'

One bell struck as Paddy finished his yarn, and I went along to rouse Gilroy out and turn in myself, much more heartened than when I came on deck.

This 'bit of a dusting,' as the Third called it, was the forerunner of a good hard blow that chased us all across the Bay. For three days we ran with only double top-sails and foresail set, while all the time the wind blew savagely true out of the nor'east and green dollops of water rolled in unceasingly over the weather rail.

Our half-deck all this while was a scene of hopeless misery. Situated in the middle of the for'ard deckhouse, it had a door on either side – ill-fitting iron affairs that even when closed let in avalanches of water. We always kept the one on the weather side closed, of course, and endeavoured to make it watertight with bits of rag and ropeyarn. But it was no good. After we had laboriously caulked it, the next big sea that came aboard would hit the door,– crash! Out all our handiwork would be washed and the icy water would spurt and cascade through the cracks. It was weary work. We gave it up after a time and grimly suffered the inconvenience of a foot or more of water swishing back and fore with every lurch of the ship.

All this while poor old Beckett lay in his bunk, groaning miserably. He was so very sick he had reached a point of absolute indifference to all mundane things – neither the wind nor the sea, nor mealtimes, nor the sodden discomfort of his surroundings troubled him any more. After a time we gave up offering him advice and encouragement and let him fight the matter out with his constitution.

The rest of us got on better: we escaped sea-sickness and did not miss a watch on deck. Jimmy, it is true, did feel qualmish once, but the old man told him with such terrible vehemence that he 'wasn't going to have any sick boys in *his* ship,' that Jimmy bucked up and tried to forget it.

The food, too, soon grew more palatable to our ravenous appetites. We quickly finished the delicacies (Beckett's included) that we had brought away with us and were faced with the alternative of eating what was provided or subsisting without material nourishment at all. We chose the former course, and before we were across the Bay were as keen about getting our 'whack' as the oldest sea-lawyer in the foc'sle. Certainly it was not much: one meal a day, and one only, was our portion. That was dinner, which consisted of a plate of pea-soup and three quarters of a pound of pork, or a plate of bean soup and a pound of beef, on alternate days. For the rest we had a pannikin of hot liquid, called 'coffee,' for breakfast, and a similar pannikin, described as 'tea' for supper; with sea-biscuits ad lib. That was all; and though it hardly sounds sumptuous, in reality it was far worse. The meat was weighed before cooking and included both fat and bone – in generous proportions. It was as hard as a brick and the only flavour it had arose from its degree of rottenness. For twenty-four hours before boiling it was soaked in the steep-tub – a large receptacle filled with salt water that was kept under the foc'sle-head ladder. It fell to us boys to replenish this tub every morning, and devilishly the water stank when we emptied it away after a dozen greasy lumps of pork had floated around in it overnight.

All our food, moreover, had to be carried from the galley to the half-deck, no easy matter in heavy weather to such greenhorns as we were. Many a plate of soup followed its owner into the lee scuppers, and many a piece of beef that had been chivvied along the swirling decks did we recapture and devour to the last scrap in gloomy hunger. Before we had been at sea a fortnight our only fear of a gale of wind was that it jeopardized the safety of our precious dinner.

The wind took off a little when we got down to the latitude of the Spanish coast and the topgallants'ls were set. The weather steadily cleared and soon we were bowling along under all plain sail, and the world began to look a different place. The fore-rigging was quickly a-flutter with clothes and bedding that the men had brought out to dry. We boys followed suit and a few hours of sunshine wrought a vast change in our comfort and in our outlook on life.

I was just beginning to enjoy things, and turned out for my afternoon watch on deck with seamanlike alacrity, but hardly had I got my foot over the threshold before Paddy hailed me:

'You been up aloft yet?' said he, sharply.

I admitted I hadn't.

'Well, up you go then,' said he, and obediently I swung myself into the main rigging and clambered cautiously aloft.

'Hold on to the shrouds, not the ratlines,' he shouted, 'and don't look at your feet!'

Again I did as I was told and made my way up till I was stopped by the futtock shrouds at the head of the lower rigging. They stretched outwards to the edge of the maintop, and realizing that to surmount them I should have to turn on my back and haul myself up by sheer strength of arm, I thought I had done enough for one day.

But 'Go on!' sternly encouraging, came to me from below and, taking my courage in both hands, I went. With a desperate effort I swung myself over into the top and, almost breathless, caught hold of the topmast rigging and dared to look about me.

Strange indeed everything looked! Up there one seemed in a world of canvas and cordage, lost in the wide spaces of the sky. The great bellying mainsail was actually below me and the deck looked a perilous distance beneath.

But I was not allowed to dwell for too long on the newness and wonder of the scene.

'Go ahead, get along with you,' came an impatient hail and flinging myself, spread-eagled, into the topmast rigging I went up step by step.

Interminable that ascent seemed, and when, after long ages of climbing, I reached the topmasthead, I had another shock. The complicated web of rigging that led to the comparative security of the cross-trees was worse than the futtocks, and the spidery ropes looked ridiculously flimsy and insecure. With a horrible qualm I entrusted myself to them and – hardly knowing how I managed it – found myself standing in the cross-trees.

Encouraged by this I climbed gingerly up the t'gallant rigging. It was shorter than the topmast – thank heaven! – and ended abruptly at the t'gallant mast head. Nothing remained above me but the bare pole of the royal mast, with the tie-chain abaft, and on either side, but out of reach, the port and starboard royal backstays. 'Well, this is as high as one *can* get,' thought I, as I clung grimly on and wondered what to do next.

But I was not yet at the end of my trials.

'Don't stop: go right on!' — Paddy's powerful voice sounded faintly up to me at that tremendous height.

I didn't move: I didn't see how I could: it seemed impossible to get further. .I simply hung on.

Paddy must have realized I was at the end of my resources for he jumped into the rigging and came swarming up like a cat, taking two ratlines.at a time, and swinging himself over the futtocks and crosstrees with hardly an effort. He was up beside me in no time. 'You're all right,' he said, 'now follow me.'

He didn't give me a chance to tell him I would rather not, but reached out and grasped the royal backstay with one hand, swinging his feet clear of the ratlines. For a fraction of a second he hung by one hand only, then twisted his legs round the backstay and swarmed up in a flash, ending by swinging himself over and astride of the royal-yard. 'Come on,' he said encouragingly, 'you're as safe as houses.'

I had my own doubts about that, but 'anyhow,' thought I, 'it's do or die, so here goes!' I grasped the backstay, jumped clear, hung for the briefest moment with only empty air for a hundred odd feet below me, and scrambled madly up. The Third's hand grabbed me by the collar and almost lifted me on to the yard beside him. 'You see that truck,' said he, pointing to the little ball at the top of the six feet of bare pole that formed the masthead, 'shin up and touch it!'

Almost mechanically I did so, then slid down and got back astride of the yard, firmly clutching the tie-chain with one hand. Then, and then only, did I dare

to look about me. And truly it was a wonderful
sight!

We seemed to be sensibly nearer the great blue arch
of the sky. There was nothing above, there was nothing
around, but the windy vault of heaven, save just in front
where, level with us, rose the swaying majestic spire of
the foremast. Beneath, in all directions, for endless
league on league, stretched the blue sea. Its steel-
clear edge, ringed with the curving sweep of the sky,
seemed infinitely far away. Below, in shapely tier on
tier, stretched the rigid, swelling shapes of the sails,
criss-crossed, stayed and pinioned by the orderly tracery
of the rigging. The deck of the ship was but a narrow
strip immeasurably far beneath us, a mere insignificant
splinter of wood in the immensity of the sea. No sound
from the little world below reached the unfathomed
blue calm about us.

Paddy pointed out to me the various ropes and spars
and explained their uses, then leisurely made his way
down again. I followed him, and, regaining the deck,
looked up at the dizzy heights from which I had just
escaped and wondered how often I should have to go
aloft.

But, curiously, after this first soul-searching experience it came wonderfully easy. I was awkward at first,
of course, but in two or three days had grown accustomed to it, and was ever after quite indifferent to the
mere fact of height.

Gilroy had his first experience aloft that same afternoon and the old man sent word along that Beckett was

to turn out and have a breath of fresh air. The latter
was feeling a bit better and rigged himself out in his
uniform and badge-cap to come on deck. Why on
earth he did it I do not know, unless he imagined the
outfitters' advertisements were correct and 'prentices
really sailed the seas in a glory of brass buttons, with a
telescope neatly tucked under one arm.

I took my first trick at the wheel in the dog-watch.
I found steering no easy matter, but under the ad-
monitory eye of the mate made a fair shape at it. While
there I heard a cry of 'Sail-ho!' from for'ard and, looking out, saw a fleck of white on the horizon far away
on the starboard bow. As it drew closer it turned out
to be a four-masted barque, running free, with all sail
set. She was homeward bound and, in the interest of
watching her, I got the ship far off her course and
ground the wheel up and down like a coffee-mill in my
efforts to steady her. A shout from the mate brought
me up with a round turn and glued my eyes to the compass, and before I was able to steal another glance, the
homeward-bounder was abeam, but too far off to make
out her name. Half an hour later she was out of sight.

Steadily, day by day, the weather improved. The air
grew soft and warm; the skies became open and sunny
and the sea turned to a deep flashing blue. The change
in the elements was reflected in our life on board. The
bad weather and its hardships were at once forgotten.
We no longer spent the whole of our watch on deck
attending to the sails and up to our waists in water.
Instead, all hands were turned to at pleasanter and much

less arduous work. Several of the men were employed aloft in fitting chafing-gear and repairing the wear and tear caused by our stormy setting out; while we boys were learning a multitude of new duties that interested and kept us busy.

One by one our hard-weather garments were laid aside. Salt-soaked jerseys, sea-boots and oilskins were hung up or thrown into odd corners, until our attire had been reduced to a pair of duck trousers and a cotton shirt apiece. We discarded boots and socks too and went barefoot. It felt strange at first and going aloft was painful, for the thin ratlines hurt the soles of one's feet, but we soon became inured to that. For the next month a shirt and trousers was the extent of our attire, except, of course, the inevitable sheath-knife strapped on the hip – the necessity for carrying which sailors emphasize in an unprintable proverb.

In the half-deck we brought to light again the photographs that had been hastily crammed into sea-chests and kit-bags when we encountered the first breath of the Atlantic. Various little 'gadgets' and knick-knacks presented by friends at home were brought into use; and during our watch below I knocked together, with the good-natured assistance of the carpenter, a couple of shelves to hold my little stock of books.

But the warm weather also brought into prominence a hitherto unguessed-at nuisance. The iron sides of the deckhouse were lined with wood, and the space between afforded an ideal breeding-ground for un-numbered hosts of cockroaches. As the weather grew more genial, so more and more of these little pests came out on us, until we were almost driven to desperation by their onslaught. They ate every scrap of food that was left in the locker from one mealtime to another; devoured it to the last shred and polished the plate clean. They floated, a thick scum, on the surface of every tin of tea or coffee that we got from the galley. They even gnawed the skin off the soles of our feet as we lay asleep in our bunks. Nothing that we could do seemed to diminish their numbers. The steward next door waged unceasing war on them and sacrificed holocausts, but it was all no good. The cockroaches fairly fought their way into possession, and after a time we more or less gave up the contest and shared our quarters with them. We put on socks before we turned in and mechanically blew a clear spot on the surface of our mugs before taking a drink.

We soon grew to like the food, rough though it was, and eked out our scanty fare by concocting various dishes beloved of apprentices. The chief constituent of all of them was sea-biscuit — sea-biscuit broken into pieces and baked with small morsels of beef or pork and called 'cracker-hash'; sea-biscuits soaked into a pulp with water and sugar, and known as 'dog's-body'; or — most delectable of all — sea-biscuits pounded up fine in a canvas bag, by the simple process of hammering it on the fore bitts with an iron belaying-pin, and then mixed into a thick stodgy cake with fat, sugar or molasses, and baked in a bully-beef tin. This was 'dandyfunk' and the most esteemed delicacy on our bill of fare.

But our best endeavours to augment our allowance of 'pound and pint' did not go far, and the 'apprentice's grace' was often voiced with bitter earnestness by some hungry individual prowling in the bottom of an empty mess-kid:

'Three between four of us,
Thank God there are no more of us!'

On the tenth day out and thereafter every day at noon we each received a small quantity of limejuice. It was bitter, very strong and highly appreciated. The practice of serving it out daily is compulsory on board British ships and required by the Board of Trade as a preventive against scurvy. The custom has resulted in the coining of the word 'lime-juicer' – a name used by seamen to denote any foreign-going British sailing ship and often applied by foreigners – notably the Americans – derisively.

In the evenings, that is to say, in the second dog-watch between six and eight, we began to have impromptu half-deck concerts. Gilroy, it proved, was a fine hand at the old Scotch ballads, while Paddy, who often came along, could sing any number of rattling sea-songs, and Jimmy had an accordion on which he would accompany anybody or anything. These sing-songs were very jolly affairs and in the course of the passage we managed to get quite a fair band. Always as eight bells struck our concerts came to an abrupt end – the watch below to turn in and the watch on deck to stand

49

by under poop or foc'sle head in readiness for a call. At sea the second dog-watch is the time for relaxation and amusement, and a certain latitude is allowed then which at other times would call for drastic punishment.

But these first days of fine weather were above all a revelation to us first-voyagers in the knowledge of the ship. When first I had seen her, embayed amid the cranes and warehouses of the East End, she looked grimy and unkempt from her long stay in port. So too in the bewilderment and physical weariness of our wild romp down Channel and battering across the Bay I had hardly taken in any details of my surroundings.

But now I looked about me and found I was in a different world. The ship was the same, yet utterly different. She was alive now instead of dead; clean, scrubbed and burnished; neat and orderly aloft and alow, where before had been a disreputable confusion. The stately fabric of her seemed to possess a life of its own. She leaned or dipped to every gust of the breeze and every surge and undulation of the sea. There was a place for everything and everything was in its place.

Now, too, she was peopled by a little company who understood and skilfully tended her. On the poop was always a controlling intelligence in the shape of the captain or officer of the watch, and at the wheel leaned the alert figure of the helmsman. Amidships the galley funnel smoked and Tommy bustled about with much clattering of pans and superabundance of energy. And on deck and aloft the men moved and worked and lent a familiar air of purposeful activity.

And how glorious, I often thought, the barque looked as she curtseyed her way to the southward! To stand at the break of the poop and, looking for'ard, watch the sweep of her bows as they rose and fell against the deep indigo of the sky was a sheer delight. Or, again, to glance aloft and see the great swaying pyramids of snow-white canvas towering up into the blue was to behold a picture of insurpassable beauty.

On the eleventh day out we were in the latitude of the Straits of Gibraltar. The breeze freshened in the forenoon and hauled ahead and just at midday, as we of the port watch were going below, the royals were clewed up.

Eight bells having struck, one of the starbowlines swung himself into the rigging to lay aloft and furl the sail, but the old man, who was standing by the poop rail, called out to the mate:

'Send one of the boys up, Mr. Thomas.'

Anxious to show I had mastered the art I jumped into the rigging and swarmed aloft. Over the futtocks I went, managed the cross-trees all right and laid out on the royal yard. The sail was bigger than I imagined and looked a much more formidable task than it appeared from the deck. The footropes swayed awkwardly, there seemed nothing much to hold on to, and though the buntlines were close up, the sail was thrashing and banging about in fine fashion. Moreover, the gentle roll of the ship, which was nothing much when felt on deck, caused the royal yard, a hundred and twenty feet in air, to sway from side to side in a wide sweep, moving swiftly across a big arc of the sky. However, there I was, and as I knew enough to pick up the bunt first I proceeded to tackle it. Digging my fingers into the canvas I dragged it in inch by inch and, after much effort, seized the gasket, hauled it tight and made it fast to the shackle of the tie.

Then gingerly I edged my way out to the weather yard-arm. The sail was blowing back over the yard and made it necessary to keep a firm hold of the jackstay and be ready to duck, to avoid being hit in the face by the slatting canvas and knocked backward. But I got out, grabbed hold of the foot of the sail and began a long tussle to master it.

I do not know how many times I dragged the canvas on to the yard before I succeeded in making it fast. Sometimes I would get it there and be unable to hold it, sometimes a gasket wasn't within reach, and each time the canvas blew out and all was to do over again. But at last I got the gaskets passed and the sail lashed to the yard – somehow.

Breathless but cheered, I scrambled over to leeward and began a similar struggle on that side. It was easier than the other and in something less than half an hour I had the sail stowed and all well fast. So I swung myself into the rigging and clambered down on deck.

By this time it was, of course, well into my watch below and I was hastening to the half-deck when the old man, who was still on the poop, called me:

'Well, Clements, do you think you can stow a royal now?' he demanded.

I replied confidently that I could.

'Then up you go again and do it,' said he, 'and' – raising his voice – 'make a better job of it this time!'

With less enthusiasm than before I turned again to the rigging, climbed aloft and, laying out on the yard, cast adrift those hardly-won gaskets. I was hesitating whether I need let go the bunt-gasket too, when a loud hail came floating up:

'Leggo that bunt! and look lively now!'

Realizing there was no help for it, I let the end fly. The sail ballooned gracefully out and there was the whole thing all adrift, tugging and leaping like a crazy bladder.

Doggedly I turned to; mastered it bit by bit all over again and finally got it fast – a little more quickly and neatly than the first time, I think.

I hurried down again as quickly as I could, for the watch was slipping by and my dinner – or what the others had left of it – was growing cold all this time. I had just reached the half-deck door, hungrily expectant, when the old man hailed me again:

'Lay aft here' – his voice had tremendous carrying power.

I turned round.

'Do you call that a good stow?' he shouted.

'Pretty good, sir,' said I dubiously, wondering what was coming next.

'You do, eh? Damme if I do! Jump aloft and furl it again.'

Bottling my indignation well down I mounted the

rigging, got on the yard, cast loose that infernal sail and painstakingly re-stowed it all over again. I had hardly got my foot on to the t'gallant rigging before I found that the old man was still not satisfied.

'Royal yard there! cast that sail loose,' — there was no pretending I didn't hear that skysail-yard voice of his.

With a feeling of despair I climbed out once more and went through the whole wearisome job afresh. At the end of it came another hail — I felt like going down and standing the consequences, but decided I had better stick it.

Again and again at an order from the captain I flung the canvas adrift and re-stowed it and stowed it again. Very soon my back was aching, my fingers were raw and bleeding, and I was drenched with perspiration and buffeted all over with the thrashing canvas. But that didn't save me. For four solid hours I was kept at it on the yard. Not till I had made the sail fast for at least the tenth time, and one bell had been struck to call the watch — my watch — was the old autocrat on the poop satisfied.

When at last I did climb down the rigging I was just about dead-beat and every limb in my body seemed dislocated. Eight bells struck as I stepped on deck. There was no help for it, my dinner was a thing of the past, and without even a chance of running below for a smoke I turned to with the watch at washing down.

It was the first lesson in practical seamanship that I received from Captain West and I quickly learnt that it was typical of his instruction, on one side at least.

Even now it appears to me a somewhat drastic method
of training but, after all, it met with a fair amount of
success. I know this particular instance, anyhow,
taught me thoroughly and unforgettably how to stow a
royal. I never needed any more teaching on that point.

CHAPTER THREE
Trades and Tropics
*

As we were now drawing near the region of the steady 'trade-winds' and the weather had become quite settled, the opportunity was taken to 'shift sail.' A ship, unlike the majority of her sex ashore, always wears her best clothes in bad weather, and when skies are soft and winds gentle, changes into her oldest and most worn attire. Appearances have nothing to do with the choice; it is a matter that the winds decide, for only the newest and strongest canvas can sustain the winter gales of the North Atlantic or the world-encircling sweep of the great winds of the Southern Ocean.

Early, then, one morning all hands were roused out to make the necessary change. The suit it was proposed to bend was brought on deck, and gantlines were rove at the mastheads in readiness. As four bells struck all hands tumbled out, the order was given 'Port watch for'ard; starboard to the main,' and the work began.

First came the two courses. These were unbent, lowered on deck, rolled up and put away. Those that were to take their places were stretched out, bent on to the buntlines and hoisted aloft, where they were secured to the jackstay and, the sheets being shackled on and hauled aft, speedily set.

After the courses and a break for breakfast came the lower tops'ls — those aloft unbent, the fresh ones bent and promptly sheeted home. Following these, the upper tops'ls and then the t'gallants'ls. The work early developed, as no doubt the old man intended it should, into keen rivalry between the watches and an eager race as to which should finish their mast first. By the time the t'gallants'ls were being swayed aloft all hands were working schooner-rigged, going at it with their blood up. The royals went aloft to the hand-over-hand shanty —

> 'Way-ay! and up she rises,
> Way-ay! and up she rises,
> Way-ay! and up she rises,
> Early in the morning.'

The word 'morning' was hardly reached before the sails were snatched out along the yard, the sheets whipped on and a triumphant hail 'All ready, hoist awa-a-ay!' sent ringing to the deck.

It was a well-fought fight and at the end of the eight hours it lasted was only won by the mate's watch by a matter of minutes — thanks largely to the bull-strength of Mac and the fiery activity of Paddy.

After a belated dinner the Second Mate's men tackled the mizen, while we of the port watch went for'ard to the jibs and thereafter, by twos and threes, to the stays'ls. It had turned three bells in the first dog-watch before the last sail was bent and set, the old suit bundled below into the locker and all hands trooped off to their tea.

Beyond a call to relieve the wheel, no order came

from aft till eight bells were struck and we mustered for
the roll-call and the routine of the night-watch. There
was no sing-song in the half-deck that evening though;
we were too tired to do more than smoke our pipes and
contentedly contemplate the day's work.

All this time the fresh breeze blew and slowly hauled
into the nor'-east. On the 24th of January we entered
the tropics, with the wind a steady whole-sail breeze
and the horizon piled with fleecy masses of white cloud.
There was little doubt that we had got the 'trades' at
last. Two days later when we were in the latitude of
the Cape de Verde islands, having run 500 miles in the
forty-eight hours, there wasn't a doubt of it. The
steady, singing breeze, true to a handsbreadth; the
tumbled whites and blues of the sea all a-sparkle in the
sunshine; the untroubled rim of the far horizon —
these were signs unmistakable of the authentic Trades,
the glorious winds that make seafaring a pleasure cruise.

No need to touch tack or sheet now; the ship sailed
herself and the helmsman had an easy time, a spoke or
two now and then was all she needed. The days began
to be a sheer delight. Most of the time we were
employed aloft — serving, parcelling, repairing, renewing, attending to the hundred and one jobs that a
sailing-ship is constantly in need of and that suggest
comparison of her with a lady's watch — always out of
repair. There is nothing more delightful imaginable
than sitting astride a gently-swaying yard, high above
ship and ocean, with a pot of tar and a ball of spunyarn,
for long hours on end, casting an occasional glance at

the acre of foam under the bows or the long white wake astern that tells of a good passage to be made.

Running the Trades down in a sailing-ship is an experience worth having – a taste of sea-life at its best. No words can well convey the beauty, the freedom, the glorious exhilaration of it all. The long halcyon hours wing by, with blue sea beneath and blue sky above and unending glitter of spray and sunshine. The ocean is all one's own or shared with a white seabird or two. The day seems twice as long as usual, and when the western sky floods with rainbow hues and the day's glory of gold turns to the silver splendour of night, there comes never an abatement or alteration in the steady onrush of the wind.

It would be hard to find a more beautiful sight than that of a sailing-ship running through the Trades as she appears, when viewed on a moonlit night, from the end of her own bowsprit.

Perched out there one seems alone in space, projected into a world of emptiness and utter beauty. Ahead the shoreless sea stretches dimly out into a mysterious eternity. The dark vault of the sky is powdered with stars, save where the sailing moon rides high and pours on the quiet sea a river of white radiance, a pathway to unimaginable realms of faerie.

And how lovely the swaying vision of the ship herself! In grace of outline and harmony with sea and sky she seems none other than the winged Spirit of the Night. Beneath one's perch the sharp stem shores unendingly on through the dark tumbling water, smashing it into splinters of white foam; while aloft, rising tier on tier in contours of unmatchable symmetry, a mighty cone of canvas, carved in rigid ivory or blackest ebony, sways silently amid the stars.

With the coming of the Trades we apprentices had more time to look about us and we began to take an interest in the sea itself. Frequently in the dog-watch the old man called one or other of us up on the poop and talked to us about our profession and duties. On such occasions he always used to emphasize the importance of learning to know the face of the sky and sea and studying the ways of birds and fishes. He himself was no great reader, but his knowledge of winds, weather and natural history was truly wonderful. He navigated the ship more, I verily believe, by the look of the sea, the flight of birds, the appearance of clouds and a certain indefinable sixth sense emanating from his vast knowledge of all these, than by the more orthodox method of scientific instruments. I never grasped half what he told me about such things as the movements of upper and lower clouds and the behaviour of birds, but he was extraordinarily interesting to listen to and talked to us 'boys' in a very kind and fatherly way when there was no work to be done. When there was, we had to jump like steel wires.

Captain West was a typical seaman of the old school — rough, masterful and equal to all emergencies. Steamers he held in a large-hearted contempt, but the sea he knew and loved. His handling of the ship in heavy weather or narrow waters was an education to watch, and all the time I sailed with him I never knew him make a bad landfall, or be out in his rough and ready reckoning more than a handsbreadth from the truth. Some of his methods would give a modern liner-officer cold shudders. Once when I was taking an azimuth he swept me and the instrument aside with a lordly gesture. 'A handful of degrees don't matter, my boy,' he said, 'take the sun like this' – laying a great hand edgeways on the compass and squinting along it at the sun – 'and keep your eyes about you.'

In one way or another our commander played a large part in our lives. All the time I knew him his was always the dominant personality on board. And he influenced very considerably the apprentices who sailed under his command. From the mate downwards we all stood more than a little in awe of him, but it was an awe mixed with liking, and when other ships and other men were in company we took pride in telling of his skill and masterfulness.

But to hark back to the Trades. One of the first denizens of deep water, and one of the most interesting that we saw as soon as we got down to warmer latitudes, was flying-fish. Just a few at first, then shoals and shoals of them. They simply swarm all over the tropic seas, and though they are preyed upon by numberless enemies – bonito, coryphene and albicore – their numbers are almost infinite. 'Flying-fish weather' sailors often call the latitudes they inhabit. Their so-called wings are thin gauzy affairs, bearing no resemblance to those of a bird but more like a huge dragonfly's. Their flight is really no more than skimming; they rise out of the water with a flip of their tails and vibrate their wings like planes. A couple of hundred yards is about their limit, then back they fall into the water with a 'zip' and, a moment after, leap out again and are off on another flight. I have often heard it said that they can fly only as long as their wings are wet; whether this is so I do not know, but it is certain that they only remain in the air ten or twenty seconds and can do little more, I should think, than throw their pursuers momentarily off their track.

Another and a very different species of deep water-man that we met with were whales. I well remember the first that I saw. It was one evening, off the Canaries. I was looking out over the rail, waiting for two bells to strike and tea to be served out, when I saw, quite close to the ship, one – two – three – up to half a score, huge black bodies heave themselves silently half-out of the sea. As they rose, thin jets of watery vapour were thrown high in the air with a long hissing noise. They seemed in no hurry, but with an easy gliding roll that hardly rippled the water went curving under. A moment or two later up they came again – one saw the glistening curve of each huge bulk – then down once more they plunged. They were most fascinating to watch, in movements so vast and leisurely, yet they did not ruffle the surface so much as a swimming man would have done.

Ever as we sagged south, the sun's rays increased in intensity and, as we drew near the Equator, the good

62

breeze began to fine away. It carried us to about the fifth parallel of North latitude and then vanished altogether. The midday sun flamed down on us almost straight overhead, the sea took on an oily, sluggish appearance, and, after a few aimless and wandering gusts, the breeze died away and we were left becalmed — the centre of a clock calm. The Doldrums had us in their grip.

'There passed a weary time.' We lay helpless on an oily sea, the spars sticking up idly into the still air. The heat grew terrific; bare iron was too hot to touch and the pitch bubbled out of the seams between the planking and stuck to our bare feet. From the yards the sails hung listlessly in heavy folds; we hauled the uselessly-flapping mains'l up, and listened to the grind and rasp of the parrals as she rocked to some imperceptible underrunning swell.

It was exasperating. The man at the wheel leaned idly over the spokes, for she had lost steerage way; the half-deck was like a furnace and our only amusement was scanning the face of the brazen heavens and whistling for a wind.

At times the sky would become overcast, a black thunder-cloud overspreading everything and then, without a moment's warning, down came the deluge! The clouds seemed to collapse, the rain fell in sheets and torrents, with such intensity that it was impossible to see more than a few yards, and the sea gave off a continuous hissing roar under the violence of the downpour.

Then, as suddenly as it began, the rain stopped, the

63

clouds rolled away and the sun shone down again with undiminished vehemence upon our drenched and steaming barque.

Sometimes at the approach of one of these squalls we stopped up the scupper-holes, and in a few minutes the main deck was inches deep in water. All hands would rush out, stark naked, with soap and dirty clothes and indulge in an impromptu bath and washing day for the quarter of an hour or so that the deluge lasted.

The vagaries of the wind, too, were heartbreaking. For the greater part of the time no breath of air stirred to quiver the dull folds of canvas or cool our heated bodies. If a match were struck its flame burnt steadily upward without a flicker. Then the merest suggestion of a breeze would whisper from ahead and round the heavy yards would be dragged to take advantage of it. No sooner had they been braced up than, as if in mockery of our efforts, a faint puff of air would come from the opposite quarter. Savagely the officer would give the order 'Weather fore brace!' and savagely the watch would throw down the freshly-coiled gear and haul the yards round to the new tack.

So it went on, day and night; the ship making a bare mile of headway in the twenty-four hours and everybody's temper frayed to breaking-point.

But even in these regions hated of the sailor we had some little excitement, for several sharks appeared. The old man first noticed one lazily swimming under the counter and — keen sportsman that he was — sent me to fetch the shark hook. This was a fearsome-looking

64

implement, fully a foot in length and sharply barbed and pointed; a swivel and several feet of chain were attached to it to avoid turns and prevent the shark severing it with his teeth.

Seeing what was afoot, Tommy came running along with a lump of pork. This was securely lashed to the hook with wire, and a coil of stout rattlin-line was bent on to the chain. When all was ready the end was made fast and the old man threw the hook overboard.

The shark was then swimming on the far side of the counter. He seemed to scent food, for he turned in his tracks and swam leisurely toward the bait. As he drew abreast of it he half-turned on his side. We saw the white gleam of his belly as, without an attempt at finesse, he made a swift snap at the pork. It disappeared, there was a violent tug on the line, and a shout from the old man. Beckett and I dashed to bear a hand and several of the men, who were peeping expectantly over the poop ladder, ran up. The shark was soon mastered and dragged, jerking frantically, up to the rail and inboard. He was not a very big one, only about five feet in length, and a few heavy blows on the head with the carpenter's maul settled him.

Hardly was he despatched before Beckett gave a shout and pointed astern. Following the direction of his hand, we saw the triangular fin of another shark gliding through the water well out on the quarter. As quickly as we could we re-baited the hook and dropped it overboard again. For a few minutes it trailed there, seemingly without attracting the attention of the shark.

Then suddenly the cruising fin disappeared, but before we had time to feel disappointed, we saw a monstrous blue shadow glide under the counter a few feet from the bait. It was a monster this time and no mistake.

The shark seemed to take no notice of the morsel we had provided for him. He swept past in a half-circle, swimming about six feet deep. Almost out of sight he went, but came back. He swam slowly round for a few minutes, then, giving a flick with his tail as though he were bound somewhere and it was time to forge ahead, he suddenly turned, made a swift lunge upward at the pork, and was hooked.

Half a dozen of us flung ourselves on the line and hauled for all we were worth. It was brand-new, twenty-one-strand hemp, but as we felt the strain, we thought it would never hold him. Fighting desperately, we managed to drag him just clear of the water and, as we did so, with a last furious convulsion he lashed himself free.

'Come up the line!' yelled the old man and, as we dropped it, flung the hook back into the water.

Then we had an example of the shark's voracity.

The hook had hardly touched the water before the great brute leapt at it again, biting so furiously that he almost swallowed the pork and the barb came through halfway down his throat. There was no doubt about it this time. Putting all our strength into it, we dragged his head well above water. He was too big a fellow to haul aboard in the same way as our previous capture, so a running bowline was sent down the line, shaken over his body and hauled taut close to his tail. The bight was then passed through a snatch-block made fast to the boom-end and we hauled him up horizontally. When high enough, he was swung inboard with lines to head and tail, dragged to the break of the poop and dropped down to the lower deck with a crash.

He was a good ten feet long and lashed about so furiously that he completely took charge. For a few minutes we couldn't get near him. Then Stedman managed to ram a windlass bar down his throat as the rest of us hung on to the lines that were fast to either end of him, and Chips, sliding a hatch under his after part, cut off his tail with an axe.

After that we set to work to cut both fish up. The tail of the bigger one was triumphantly carried for'ard and nailed to the extremity of the bowsprit, the longer lobe uppermost. This is always the custom when a bigger shark than usual is caught, and constitutes the time-honoured decoration of a deep-waterman.

Next day we caught two more sharks, both about six feet long. The men called them 'shovel-nosed' sharks and I cut out the jaw of one, with its triple rows of triangular knife-edged teeth, to keep as a curiosity.

For a whole week we lay drifting about on a windless sea. Save for a few minutes at a time, when a catspaw ruffled the surface of the water, we were without steerage way. The excitement afforded by the sharks – we caught eight during the week – was a welcome relief to the wearisome monotony of hauling the creaking yards round to every puff of air.

67

One morning the skipper spied a dolphin gliding and darting through the water on the port side of the poop. He sent for the grains, a weapon like a double trident with five barbed points, fitted to a slender wooden haft and having a line attached. Then he climbed out on the bumpkin and, watching his opportunity, drove the grains down just as the dolphin flashed beneath him. He speared it the very first shot, a piece of skill worthy of our Nimrod of a skipper.

We quickly hauled the dolphin aboard and a glorious fish he was, nearly five feet in length, with scintillating metallic hues. He looked superb gliding through the water like a flash of many-coloured light, but his brilliancy soon dimmed as he lay quivering on our sun-scorched deck. We carried him for'ard and cut him up, and cabin, foc'sle and half-deck had a sumptuous tea, for the flesh of the dolphin is white, dry and very good eating.

The many kinds of fish we saw caused the old man to overhaul his gear, and very soon the poop was decorated with the implements of the deep-sea fisherman's craft. On one side, slung in beckets, was a heavy two-flanged harpoon fitted to an eight-foot shaft, with a thick line attached and almost strong enough to hold a whale. On the other side was a smaller harpoon with a single barb that drew across in the shape of a T, a deadly weapon of the type used by the Yankee whalemen of Nantucket and New Bedford. Underneath were the grains and coiled near the rail the shark hook and line. All were ready for instant use and the old man never let slip an opportunity of bringing them into action.

All these things helped to wile away the tedium of the doldrums. And then, at last, to our relief, wind came. It was at sunset on the eighth day of windless drift that we saw, well out on the port bow, a dark line on the water and felt the merest suggestion of movement stir the heavy air. The dark line drew closer and, as it did so, resolved itself into a ruffled patch on the surface of the greasy sea. The old man was watching it intently. A few moments satisfied him of what he saw, for, coming to the poop-rail, he gave the order 'Starboard fore brace' in a tone that betokened he anticipated a change of some sort.

The little ripples touched the ship's side and a cool breath of air fanned our cheeks. At the order we sprang to the braces and hauled cheerily.

The heavy yards swung round, the sails flapped and filled and the barque began to make way through the water.

The welcome breeze kept on blowing gently, but steadily. The main sheet was hauled aft and the hitherto useless stays'ls set. The gear was coiled down and still the gentle breeze blew and the cheerful sound of rippling water arose.

At eight bells the wind still held and the ship was logging an easy five knots. As the wheel was relieved and we turned away for'ard, Mac voiced the general opinion –

'Lay your shore-gear handy, me lucky boys,' said he, 'the port watch ha' brought ye the Trades.'

And so it proved; all night the breeze held, blowing steadily from the sou'-sou'-east and imperceptibly increasing in force. It was about 1° North latitude where we picked them up and next morning, the 14th of February, word went round that we should be crossing the Line that day.

The time-honoured ceremony of crossing the Line has been so often described that I do not propose to inflict an account of it upon the reader here. In its main features the procedure was the same on all ships, and only differed in details according to circumstances. Invariably it began with much laughter and practical joking, culminated in an orgy of Stockholm tar and salt water, on the part of the unfortunate novitiates, and ended in a bottle of whisky being sent for'ard to the men and in the singing of many shanties by all hands.

One might call it rough humour, but it is a quaint old custom nevertheless, and one that only takes place on a contented ship where all hands pull well together. It is 'excellent fooling' and does much to enliven the routine of a long passage. All these old sea-customs are rapidly dying out, more's the pity. They are passing with the passing of Sail and soon will be but a memory.

After crossing the Equator, for a fortnight we bowled steadily along through the south-east trades. Day after day, with all sail set and yards braced up on the port tack just clear of the backstays, we surged south over summer seas and under soft blue skies. All hands were employed aloft, overhauling the rigging and reeving new running gear in readiness for the great seas and gales of the 'roaring forties.' In the watch below the older men, with the wisdom of experience, mended sea-boots and oilskins against the bitter days ahead; even the steward rigged up a contrivance of wood and canvas to fit over his weatherdoor and save his leaky galley from being washed out; but we boys fleeted the time carelessly as they did in the golden world. We spent our leisure in reading, sing-songs and boxing – the latter under the tuition of the Third Mate, who brought along a pair of old gloves and under the guise of instruction knocked us about most unmercifully. But we progressed and our whole-hearted set-to's were sometimes conducted under the patronizing eye of the captain himself.

So the days passed in fair summer contentment. The present sufficed us, but the writing on the wall was not wanting. 'Just you wait,' the Second Mate used to say, 'just you wait, my bold young sailormen, till we're off the pitch of the Cape!'

One night we passed a large homeward-bound barque. It was bright moonlight and she swept past a couple of cables' length away. Very beautiful but ghostly she looked, swaying silently by, with the moon-shine gleaming on her snowy canvas and the pale green of her starboard side-light glimmering like a spectral eye. For a few moments she was visible and then, ghostlike, was swallowed up by the night.

After a week of such ideal sailing we sighted one morning the little island of South Trinidad, far out to sea off the Brazilian coast. At the cry of 'Land-ho!' I ran out on deck and climbed into the foretop. Fine on the starboard bow it lay, a small cloudy speck, motionless on the horizon. It was the first land we had seen since leaving England and the sight of it held me fascinated.

Some hours later we passed quite close to it and a more rugged and romantic spot I never saw. It is very lofty, with bare, precipitous cliffs rising sheer out of the sea — split into deep ravines, scarred and jagged with black, fantastic rocks and ringed with a thunderous line of surf. Far out to sea on the port hand we could just discern a broken cluster of rocks — the Martin Vaz Islands.

As though Trinidad stood sentinel at the portals of sterner seas, the weather gradually changed. As soon as the island was astern the air grew more chilly and the nights turned downright cold. Day after day the sky assumed a harder tinge, the wind came more and more in gusts, and the frequent sprays that leapt through the fore-rigging struck one's face with a vicious sting. Slowly we laid aside our thin tropical gear and donned warmer garments. The flying fish disappeared, our hunting tackle was put away and almost insensibly we reached the boisterous realms of the great West Wind.

CHAPTER FOUR
The Pitch of the Cape
*

THEN suddenly, on the night of Sunday, March 1st, the weather broke up. All day the wind had been freshening and when we of the port watch came on deck at midnight the prospect looked threatening. The ship was still carrying every stitch of canvas, but at times, to a fresher puff than usual, she staggered under the tremendous pressure and a long, white cascade foamed aboard over the almost submerged lee bulwarks. The mate, who hadn't spent his life in the tea clippers for nothing, wedged himself into the corner of the poop-rail and let the barque rip. And rip she did. Paddy, Beckett and I stood ready under the break, momentarily expecting a call. But Mr. Thomas hung on to his canvas. One bell was struck; two bells; a moment later a few sharp drops of rain began to patter on the deck and the wind's note shrilled to a keener edge.

'Stand by your stays'l halyards!' came the order, and then – to the rising drone of the gust – 'Leggo! Aft here, and brail in the spanker!'

'Brail in the spanker!' shouted Paddy, springing to the pin-rail and throwing off the coils, as the men came blundering along from for'ard.

'Lee brails, some of you! – Clear that head-outhaul! – Leggo your gaff-tops'l halyards!' The squall shrilled fiercely, and under the strain the barque hummed like a top aloft.

73   c*

Cursing the darkness and the beating rain, the men searched for the ropes and with quick cries began to spill the wind from the shaking canvas. Hardly was the sail snugged in to the mast before there was a sharp crack for'ard, a quick rattle of chain and a noise like a volley of pistol shots.

'There goes the fore-royal,' said someone.

'Let go your royal halyards,' shouted the mate, and we dashed for'ard to carry out the order. The yard came down with a run and as it did so, bang went the fore-t'gallant sheet. For a moment the sail blew out, tugging wildly at the bolt-ropes, then split into ribbons and a thousand tatters of canvas whirled away into the darkness.

'T'gallant halyards!' bellowed Paddy, 'damn it all, what are ye standin' lookin' at?'

A man sprang to the halyards and as the fall whirred through the sheaves there was another roar, a quick rattle of hanks in the bows, and we heard the look-out man's voice: 'Inner jib's blown away, sir!'

Things were getting lively and a tumbling green sea, rearing above the weather bulwarks, hung for a moment, then crashed aboard and filled the decks waist-high in a welter of foam.

'Call all hands,' said the mate quietly, 'and you, Beckett, rouse the carpenter and sailmaker out.'

A man ran to the foc'sle door, flung it open and shouted into the dim slumbrous warmth of the interior:

'All hands on deck to shorten sa-a-ail! Turn out, me sons, she's in a ruddy mess aloft.'

An uneasy stir of tired men answered him; the hands rolled out of their bunks, reaching hastily for coats and boots as they made for the door.

I ran aft to waken the second mate. He bounced out of his bunk and was pulling on his trousers almost before I had delivered the order. 'All hands on deck' is the ordinary call and permits leisurely dressing; the addition 'to shorten sail!' is the emergency one and demands instant action.

As I dashed for'ard again, with Mr. Miller close behind, cursing and buckling his belt, both watches were clewing up the t'gallants'ls. The buntlines were no sooner snugged up than there came an order to furl the mains'l and, glancing aft, we saw that the old man had come on deck.

This looked as though it would prove a ticklish job. The wind was piping up in great gusts, green seas were pouring aboard and darkness and driving rain enveloped everything.

Mr. Miller eased the sheet and the mate clawed his way over to the tack, ready to let go as soon as clew-garnet and buntlines were manned.

'All ready, sir,' sang out Paddy, as soon as the hands were at their places.

'Leggo, then,' said the mate, slipping the tack, 'haul away!'

The tack whipped round the chess-trees, the great sail bellied out, thrashing heavily, and all hands hauled furiously at the spilling-lines, heedless of everything except mastering the sail before anything carried away.

75

It was too late: we heard a sharp warning – 'Look out there! hang on everybody!' and cast an instant's glance to wind'ard. Right above the bulwarks towered a great foam-crested sea. Even as we looked, it fell, sweeping across the deck with the volume and impetuosity of a burst dam. All hands were thrown off their feet and washed into the scuppers. Clew-garnet and buntlines were torn from our grasp and before we could regain them the mischief was done. The great sail took charge. With a volley of reports like a battle-ship's broadside, the canvas soared upwards like a balloon, thrashed madly for a moment, then, with a swift ripping and rending, split from earring to earring a foot below the jackstay and collapsed bodily into the sea to leeward.

For a few minutes our hands were full. Gilroy and I were sent to cut the boltrope and fluttering strips of canvas from the jackstay, while the men, up to their necks in water, climbed on the lee-rail and by main strength dragged the sail back on board. We carried the tattered remnants aft and then systematically started to snug the ship down.

For hours we were at it. The t'gallants'ls were clewed up and when the ship was sufficiently shortened down we had to lay aloft and furl everything.

It was five o'clock and near daybreak when the last man was down from aloft and the order was given to relieve the wheel and look-out. By this time it was our watch below and we turned into our bunks in our wet clothes to snatch a couple of hours sleep.

Coming on deck at eight bells we found a change

indeed. The wind was blowing harder than ever and the great grey seas rushed past, streaked with foam and indescribably wild and grand in appearance. The ship was plunging heavily, under only tops'ls, lower stays'ls and fores'l. The canvas aloft was still our fine weather suit and was standing the strain badly. The foresail had split in two places and Neilsen and I were sent aloft with palms and needles to try and stay the rents from spreading further. We might as well have taken sticking-plaster with us. The rents soon stretched from head to foot and the old man decided to furl the sail. Before tack or sheet was started, however, two other seams split and the job looked pretty hopeless. Still, it had to be done.

All hands were called, the clew-garnets manned, the tack slipped and we started to haul away. The tack was barely a third of the way up when the sail gave two or three flutters and split literally into ribbons. In an instant we were caught in a tornado of lashing rope and canvas. Something hit me on the side of the head and knocked me flying. I lay for a few moments helpless under the crab-winch, with my head going round like a top. Stedman came and picked me up; he was bleeding himself and several others had received cuts from the flying ropes.

It took us a couple of hours to lash the remnants of the sail to the yard and then the order was given to unbend the top'sls and get up storm canvas in their places. Till six bells in the afternoon the job took us. Then, dinnerless, but with good 'number nought' canvas above our heads, the watch was sent below.

Again it was our turn and the unfortunates in the starboard watch remained on deck to clear up. They got altogether the worst of this first touch of bad weather, for they had been on deck a full twenty out of the last twenty-four hours.

The wind moderated a little during the night and we set the main t'gallants'l, but the air had turned bitterly cold and I was glad to bend all my winter clothing.

This blow was our first indication of what was coming and reminded us that the roaring forties lay ahead. The summer seas of the trades were done with and we could look forward to a battering off the Cape and a long run through the Great Southern Ocean, to the accompaniment of gales and heavy seas.

As usual, our leaky old half-deck got its fair share of the water that came aboard. The floor was awash and everything we possessed was soaked. Wet through it remained for three solid weeks. During that time there wasn't a dry patch in the place. Fifteen inches deep, to the top of the threshold, the water swirled and splashed with every roll of the ship. Clothes, sea-chests, bedding — all were wringing wet. We turned into our bunks all standing, to wake four hours later with garments moist, warm and steaming. The sensation of turning suddenly out into the raw, bitter air was horrible and like the stabbing of innumerable spear-points. Yet no one ever caught cold. Colds are unknown at sea, the salt water and salt-impregnated air prevent continual drenchings from being more than merely unpleasant.

Now, too, the scantiness of our food became more of a hardship. In the tropics we had often been able to supplement our fare with a dish of flying-fish, bonito, or even shark steaks, but now, with the air keener and our appetites more ravenous, we were straitly confined to pound and pint. Stedman used grimly to tell us that the regulation breakfast in a lime-juicer was two draws of one's pipe and reefing another hole in one's belt.

A step-motherly Board of Trade insists that her apprentices and merchant seamen shall receive 'full and plenty.' It constituted a difficulty for economically-minded shipowners, but they overcame it. They did so by providing sea-biscuits, any number of them, which might be had for the asking. These biscuits, or 'crackers' as apprentices affectionately called them, were hard as bricks, tasteless as leather and infested with maggots and weevils. Precious few of them made 'full and plenty.'

The breaking-up of the weather and the fact that we were nearly two months out and each of the men had a few pounds due to him, brought to the fore another deep-sea institution. This was the slop chest, which was provided, not by the owners, but by a far-sighted commander who knew the improvidence of foremast Jack and laid in a stock of clothing and other necessaries with which to provide him – at a price.

Every Saturday night the slop-chest was thrown open, and the steward attended at the pantry hatch with piles of clothing and oddments, while the men drifted along and made their purchases. Shirts and jerseys, seaboots

79

and oilskins, blankets and suits of dungarees were all largely in demand. The steward was provided with a book in which he entered against each man's name the extent of his purchases. 'Sea-price' was charged, and sea-price is often a figure which a Maltee Jew would hesitate to ask, but our commander was content with a reasonable profit and there was some relation between value and price in the articles supplied.

Tobacco was also served out at the same time, and this found good customers in the half-deck. It was American cake tobacco, put up in solid half-pound plugs – 'Buckskin' and 'Lucky Hit' – and sold to us at three shillings a pound. It served as a kind of currency on board for the primitive buying and selling that we had occasion for.

As the nights grew cold and we could no longer lie down comfortably on a coil of rope, we took up our old station at the break of the poop and resumed our long yarns with Paddy.

Sometimes when there was no work doing the mate would come down from the poop to light his pipe, and stop for a few words. He was a little, elderly, grizzled man, and, as a rule, not much given to talking. For many years he had served in the famous China clippers and when he could be opened out he was chock-full of information about remote corners of the Celestial empire. Clipper habits still lingered; he was a terror for 'carrying on' and an awful fidget – never quiet himself and always raking up some little job or other for the watch. His favourite employment for us boys was over-

hauling buntlines. Every time he went along the deck
he would give a pull at each in turn and break the stops,
then, as soon as he got for'ard, whistle for Beckett and
me with the order 'Aloft and overhaul those buntlines.'

We got tired of this habit of his after a time and
stopped them up with ropeyarns so securely that he
couldn't break them. We tried it once, but never again.
We watched him swigging on the main t'gallant buntline and chuckled, but our triumph was short-lived. He
blew his whistle.

'What have you boys been doing to these buntlines,'
he said, 'putting wire seizings on 'em?'

'No, sir!' said we, in innocent surprise.

'Well, jump aloft and clear 'em.' And aloft we had to
go and pass a knife through our frappings.

No sooner had we got back on deck that the old
villain tried the buntlines again. 'Ah,' said he, 'that's
better! 'Way aloft and overhaul these buntlines, you
boys.' And aloft we had to go and make a proper job of
it, reflecting that the old mate was distinctly one up on
us.

Now too that the sun was often obscured at mid-day
dead reckoning became of importance and another of our
duties was to heave the log every two hours, to ascertain
the ship's speed through the water, and enable her position to be worked out on the basis of course and
speed.

Beckett took the glass, I held the reel and Paddy
attended to the log-line. The latter was paid easily over
the rail while a fourteen-seconds sand-glass ran out, and

was then abruptly checked and held fast the moment the man at the glass called 'Stop!'

The number of knots that had passed out over the rail gave the ship's speed in sea-miles through the water. A sharp jerk was then given to the line to spill the water out of the logship and it was hauled in, reeled up and put away till next time.

With the change in the weather sea-birds of many kinds put in an appearance, and right across the Great Southern Ocean our course was followed by great flocks. Pre-eminent amongst them were wandering albatross, the very queen of sea-birds, with a wing-spread averaging from ten to fifteen feet.

Glorious birds these latter are, with snow-white breasts, black markings on their long, narrow wings and powerful hooked beaks. They live entirely at sea and must even sleep in the air, for they only go ashore, on one or other of the rocky, desolate islands scattered about these high latitudes, during the breeding season. They seem fond of a vessel's company, or perhaps experience has taught them that such intruders usually leave a trail of scraps in their wake, for two or three are nearly always in sight from a sailing-ship's deck. When they alight on the surface of the sea in search of food they splay out their broad, webbed feet, pushing the water before them and so deaden their way.

Their powers of flight are wonderful and most fascinating to watch. They fly for hours without a movement of their wings more than an occasional, almost imperceptible, tilt that changes their elevation.

They seem, too, to have the faculty of remaining almost stationary in the air, but when they do decide to go ahead, they can pass a running clipper as though she were standing still. And the harder the gale blows, the more calm and serene is the albatross. With the wind howling across the great spaces of ocean, with never a thing to break its force between New Zealand and the Horn, and the ship under a goosewinged main-top'sl, the albatross floats in the air with never a flicker of its wings, unconcerned that the wind is rising eighty miles an hour. Head on, running, or even athwart the wind, they plane or hang motionless with never an effort. The albatross' consummate mastery of the air has never been satisfactorily explained and must, I should think, give aviation experts food for thought.

The many old superstitions in regard to these birds are dying out and we caught large numbers every time we went south of 38°. The method employed is similar to fishing, but with a difference. An ordinary fishing-line is used, carrying, instead of a hook, a piece of brass shaped like a diamond, thus ⬦ . Around the edges are sewn strips of pork fat and it is towed some fifty feet or so astern of the ship. The albatross swoops down on the pork and the hook of its beak is caught in the diamond. The fisherman at once hauls in the line, keeping it taut so that the albatross is unable to get its beak free. The bird splays out its broad feet, resenting its undignified passage through the water, and often gives its captor all that he can do to haul it aboard. The critical moment comes when the bird is directly beneath the counter and

83

just as it is lifted into the air, for if the fisherman is not very careful the line will slacken, and the albatross flap itself free. Once the bird is lifted clear of the water its fate is sealed and it is only necessary to seize it by the neck and lift it over the taffrail. It doesn't do to grab it anyhow, as Beckett did the first time he caught one. In his excitement he slackened the line, and the offended albatross gave him a peck which took a three-cornered piece of flesh out of his hand and required months to heal.

As soon as the albatross is landed on deck it becomes quite helpless and rolls and flaps about, looking very ruffled and ridiculous. It is usually seasick and disgorges a lot of undigested fish over the decks. A blow on the head kills it, and its humiliated body is put to many different uses. Sometimes the head, breast and wings are cured and mounted in one piece, making a very hand-some decoration. More often the bird is dismembered and its thick plumage and soft down plucked to make a luxurious pillow. The head is scraped and cleaned and turned into a letter-holder. The bones in the upper part of the wings are carved into paper-cutters; those in the lower into pipe stems; and the broad webbed feet have the bones removed and make handy tobacco pouches. Finally, if the bird be a young one, fillets are cut from the breast and make a second appearance in the wet-hash at supper time.

The killing of albatross always struck me as an atroci-ous deed. They are such splendid birds and the romance and mystery of their existence fascinating in the extreme.

84

Among sailors they have earned an ill-name on account of their habit of pecking at the eyes and scalp of any unfortunate floating in the water, but I believe this misdeed is more properly attributable to the giant petrels.

There are two other types of albatross, even more frequently met with than the peerless 'wanderer' — the sooty albatross and the molly-hawk. The first is somewhat smaller and not nearly such a striking bird. It is a uniform sooty brown in colour. 'Molly-hawk' is the name bestowed by sailors upon the smaller albatross; it is caught in the same way as its larger relative and we captured numbers for the sake of their plumage.

All the time we were running our 'easting down' we were followed by huge numbers of Cape pigeons and petrels. There was always a cloud of them swooping and darting astern, waiting for scraps to be thrown overboard. Such a squawking and hullabaloo there would be when the steward emptied his bucket overboard. They squabbled desperately among themselves for the choicest bits, until some lordly albatross swooped down, gave a sharp peck here and there and scattered the lot until his own regal appetite had been satisfied. Then back they would all come and squabble and gabble until not the smallest morsel was left. We used to catch them in the same way as their larger kindred, save that we used an ordinary fish-hook, which the greedy little beggars would snap at quite readily. We did it for the fun of the thing, for they were no use and we always let them fly again.

One of the men served a stormy petrel a very cruel

85

trick. He did it once, but only once – at least, while he was in the *Arethusa*. Having caught one, he tied a 7lb. bully-beef tin, with a hole punched in the bottom, to the petrel's legs and threw it overboard again. The tin prevented the petrel from flying and gradually filled and sank, dragging the little bird, struggling gamely, down with it. Paddy spotted what Johnson was doing and, as the latter hung over the rail watching the petrel drown, he ran up behind and landed him a terrific kick that made his teeth rattle. It brought Johnson down with a run and left him squirming on the deck. Paddy was much too free with fist and boot, but on this occasion at least, he used the latter to noble purpose. The darky walked stiffly for days.

On March 4th we passed close to the Tristan d' Acunha group of islands. We saw them when day broke, two or three misty little bulges on the horizon. They soon took shape as bare, rock-girt islands and when eight bells struck I went aloft to have a look at them. The one nearest to us, Nightingale Island, was small and square-topped, with some isolated pinnacles of rock off its northern end. Farther off lay Inaccessible Island—sea-battered, destitute of any blade of green and inhabited only by sea-fowl. Just for'ard of the beam lay Tristan d'Acunha itself.

It was a thick, gloomy day, with more than a suspicion of bad weather to come, and the high, precipitous cliffs, iron-bound and with a smother of foam at their base, looked wild and inhospitable to a degree. Towering above them, looming dimly through the drifting

86

patches of fog-smoke, was the great peak – the unique feature of Tristan – that rises from its lofty plateau another 6,000 feet in air. Fitting outposts of the Great Southern Ocean the islands looked, with the league-long rollers bursting on their ramparts, cut off by a thousand miles of stormy sea from the nearest land.

With Tristan astern and the barque's head still slanting to the south'ard of east the weather grew still colder. Sharp squalls rattled down on us, sleet and hail took the place of rain, but a roaring wind held steadily out of the sou'-west and under t'gallants'ls we ran romping before it.

Eight days brought us to the longitude of the Cape. As the second mate had predicted, it was a far cry from the holiday-making of the Trades. However, one expects change at sea and takes it as all in the day's work. It helped to turn us first-voyagers into sailormen; our oilskins were almost worn out, our bedding was wet and our sea-boots always full, but we chewed tobacco and jumped to an order with seamanlike alacrity.

Very glad we were to put the dreaded Cape of Storms behind us. The African mainland itself was far to the nor'ard, our latitude on the meridian of Cape Town being 43° and some minutes. The weather looked very threatening the day we got round; the long westerly swell took on a mightier swing and the wind tuned up to a higher key. In the dog-watch Stedman and Mac dropped into the half-deck for a yarn. The talk, appropriately enough, swung round to superstitions and the wild legends of the sea. The right setting such stories had that night. The confined space of the half-deck, cumbered with sea chests, and with oilskins and sea-boots a-swing on the bulkhead — the rough figures in sou'-westers and pilot jackets — the air thick with smoke that cleared fitfully to gusts of wind through cracks of the weather-door. Through the open door-way to leeward loomed the blackness of the sea, weirdly streaked with lines of foam; and over all sounded the endless noises of the labouring ship.

And the yarns — wild tales they were! Mac, as he sat there — bull-throated, bare-armed, hairy-chested — might have been none other than Kipling's *Merchant-man*, telling of the wonder and the terror of the sea as he himself had known it. An implicit believer we found he was in the stories he narrated. He said there were 'bad people' in the parts from where he came, by which he must have meant ghosts and fairies, and he gave us the story of the 'flying Dutchman' as a bald matter of fact.

'This same rampagin' ould cape is the everlastin' haunt av her,' said he. 'An ould Dutch windbag she was, with a howlin' haythin on her poop. An' for all he couldn't weather this same divil's thrap av a cape, phwat does he do but curse the blessid saints. B'me sowl! he does — he curses iviry wan av thim by name. An' for pinance it was given him to thry for iver to round the Cape an' nivir to do it. For all his baytin' 'an' baytin', the divil an inch was he to get to the east'ard. An' still he's a thryin'—bad luck to him! — and badder luck to the poor sailorman that claps eyes on his damned ould hooker. B'th' powers! they'll come to a bad end, iviry mother's son av thim, and die cursin' the day they raised the onlucky tops'ls av her.'

'I hope we don't see a ship to-night,' said Gilroy, 'or I'll feel dead sure she's the Dutchman.'

'No fear o' that,' said Stedman, 'the' Dutchman's one o' them old-fashioned rigs. Slap into the wind's eye she goes, with all sail set and the stars shinin' thro' 'em, and sounds coming from her like the howlin' o' dogs.'

'Shure they are,' echoed Mac, 'may all the saints be good to the poor sowl that hears th' noise av thim!'

Starting in this way the two regaled us with a variety of fearsome yarns. Old Chips came in later and nodded his head in grim approbation at the unfolding of each fresh horror. Stedman seemed doubtful of the tales he told, but big Mac – a true seaman and a true Irishman – believed in them wholeheartedly.

Most sailors seem to have a streak of superstition in their character. Certainly all the old-time sailing-ship men had. It is hardly to be wondered at; the lonely nature of a sea-life, spent amid wide, unpeopled spaces, lends itself to weird imaginings. The wonders of the deep, spoken of by the psalmist, is not a chance phrase. 'They that go down to the sea in ships' do sometimes see unaccountable things.

CHAPTER FIVE
The Roaring Forties
*

THE Cape, we found, did not intend to belie its reputation. The promise of bad weather was fulfilled and for a few days we had a particularly wet and weary time. Mr. Thomas said it was the result of a cyclone far to the nor'ard.

At any rate, the wind blew a full gale and was constantly hauling and veering in violent squalls of hail and rain. A nasty cross sea, running from all directions at once, rose up and green water tumbled aboard over both rails. The inky, racing clouds were so low as almost to envelop us and for forty-eight hours, even at mid-day, the air was scarcely lighter than the glimmer of a grey dawn. Lightning stabbed and flickered all round, but, strangely enough, without any accompanying thunder.

To add to our discomfort, the galley-fire was put out by an unusually big sea that curled along the whole length of our starboard rail and fell aboard in tons. Little Tommy was almost drowned among his pots and pans, but floated to the top and struggled gamely back, trying ineffectually to re-light his fire. For twenty-four hours we had only sea-biscuits to eat. We had nothing at all to drink, for we were unable to rig the fresh-water pump at the main fife-rail. We made several attempts, but the main deck was constantly flush to the rail and the salt water got down and would soon have spoilt our whole drinking supply.

Stowing the inner jib I cut my fore-finger length-ways to the bone. I plastered it well with Stockholm tar and wrapped it up in a strip of canvas – an unfailing remedy for cuts at sea – but the constant soaking with salt water made it long a-healing and gave me agonies every time a sail was made fast and I had to hand the board-like canvas.

Our wretched old half-deck was pretty well sub-merged all the time. My bunk was to leeward and almost under water. For eight or ten watches I had to sleep in Gilroy's bunk; he was in the starboard watch, so we took turn and turn about at it. As we lay down 'all-standing' – just, in fact, as we came in dripping from the deck, without even troubling to take our oilskins off – there soon wasn't much to choose in the matter of wetness between his bunk and mine.

We ran all one day and night with only lower tops'ls and foresail spread, but with the wind steadying in the south-west and signs of a break showing, we put the upper tops'ls on her. Wind and sea gradually lost their cyclonic character, and without delay we shook out the main t'gallants'l and drove 'hell-for-leather' to the east.

The fresh sou'-westerly gale held true and for the next six or seven days we did some grand sailing – lee rail under water most of the time and decks as steep as the roof of a house. It was a quartering sea and didn't bother us much, so the old man piled sail on her and she ran like a stag with the hounds on her track. Never less than ten and more often a full twelve knots she made, and when the log was hove, it was all the man holding the reel could do to prevent it being jerked out of his hand.

The rattling good days' runs we were making put everybody in a good temper and compensated for coming an occasional cropper with the soup or getting our tin of tea filled up with salt water. Even Paddy forbore to curse everybody and everything with his usual vehemence and impartiality, though he cast many a contemptuous glance aloft at the bare sticks above our straining tops'ls.

'Why don't he put the t'gallants'ls on her?' he growled. 'I tell you, Clements,' he said to me one day, 'if this ship was a "down-easter" she'd be flauntin' a main-royal. Aboard those packets they put a man on the foc'sle head to report when the cathead's in sight. He sings out as soon as he sees it shoving up out of the water to leeward, and the order comes instanter, "Give her the jib"!'

I laughed, for this little fable took some swallowing. When a ship lies over so that the cathead is submerged she is pretty well on her beam ends, and the order would more likely be 'Cut away!' than 'Loose the jib!' But I passed on Paddy's yarn and it tickled everybody. Thereafter, whenever we took a bigger sea than usual or looked like having a particularly dirty night, somebody would remark dispassionately, 'Give her the jib!'

A good part of our time, when we weren't actually aloft shaking out a t'gallants'l or picking it up again, was spent under the foc'sle head making sennit. We

92

worked ropeyarns up into four-foot rovings to lash the
heads of the sails to the jackstay and replace those that
were constantly being worn out. A 'stand-by' job was
this, one only done when the weather would not
permit of scrubbing or repairing about the decks and
the watch was kept 'standing by' in readiness for a call.
John Neilson and I were sent aft to help the sail-maker
on the poop with palm and needle. It was a pleasant
job, for one was clear of the swirling maindeck and in
comparative comfort under the lee of the weather-
dodger.

All the time we were steering more and more to the
south'ard. The captain was a firm believer in the
famous Maury's theories and shaped a composite great
circle course through the roaring forties. A typical
taste of their quality they gave us. With the 45th
parallel to the nor'ard, heavy weather set in in earnest
and we ran due east to a ceaseless accompaniment of
howling winds and bursting seas. Catching birds was
out of the question, the t'gallants'ls were fast and remained fast, and watch after watch we simply stood by
and carried out orders.

Ten days or so after rounding the Cape we passed the
lonely Crozets. They were sighted at daybreak one
morning and I first saw them when I came on deck at
eight bells. They were not far distant, and as I poked
my head out of the half-deck door I stared at them in
amazement.

Across a monstrous foam-streaked sea, through which
our little barque plunged under her ribbons of tops'ls,

a bare two miles distant rose a mighty jagged bastion of granite. Grey, naked and snow-crowned, it stood out against the steel-clear sky, the great sea thundering at its base. And, ye gods, what a sea! It was impossible to realize the power of those great rollers through which we had been running until one saw a barrier interposed to their onrush. It made one almost hold one's breath to watch some great roller, winged by a thousand miles of storm, rush towards the nearest island. With a roar plainly audible on board it hurled itself on the rocks, sending the spray flying hundreds of feet into the air. One almost looked to see the solid granite crumble and give way before such an onslaught, but as the foam fell back the rock shone out again, cold, glistening and defiant as ever. Never did I see such an awe-inspiring display of the forces of nature as here, where these ocean-strongholds oppose their ramparts to the mighty sweep of the West Wind and its attendant sea.

If the old man had come so far south in the hope of getting plenty of wind he had no cause to complain. It blew like ten furies. We got more wind than we well knew what to do with, and the night we left the Crozets astern it came on to blow in a way that the packet-rats called a 'rip-snorter.'

The mercury dropped steadily till it looked as if it would fall through the bottom of the barometer. The wind felt like pins and needles and rose in great gusts. The sleet that fell in the driving squalls was not so much snow or rain as little splinters of ice ; it filled up all the corners and froze in a solid mass that made going aloft a hazardous performance. Everything was slippery and one had to break the frozen mush on the yards with numbed fingers to get a grip on the jackstay.

The sea was a magnificent sight — an endless succession of swiftly-moving hollows and ridges. The mighty, league-long greybeards, stretching out to the horizon on either hand, swept carelessly past the ship, shouldering her a point or two off her course, first on one side and then on the other, and pouring over both rails in such cataracts that from aloft the barque looked like a half-tide rock.

All hands were kept aloft most of the watch passing extra gaskets round the sails that were stowed and lashing with rovings the heads of those still set. It was fine to be aloft above the panorama of that wind-swept, regal sea. At times one could have shouted at the wild exhilaration of it, but such moments soon ended in a shrieking squall of hail that blotted everything out, and left one only able to hang on, buffeted and gasping, with the feeling that one's ears were being cut off by red-hot knives.

The long southern night rolled down on us, with threat of still worse weather, and the barque's lurchings became more and more violent. A hand had been sent to the lee wheel, but it was obvious we were carrying more than we could stand, and at four bells the order was given to take in the main upper tops'l. The fore was already fast.

An awful job the sail gave us. The canvas was frozen and as hard as a board. It was as black as the Pit aloft, and we almost lost one another on the yard. But it was done at last and the watch sent below, the old man being grimly determined to hang on to his fores'l. Under that and the narrow band of the lower tops'ls we drove blindly all night to the eastward, the watch on deck gathered on the poop and two A.B's at the wheel.

It was just after daybreak next morning, when the starbowlines were on deck, that we of the port watch were aroused out of our bunks by a sudden shout:

'All hands on deck! Out, everybody, smartly! Main hatch has lifted.'

We rushed out, helter-skelter, in answer to the call. The main hatch is the most vulnerable part of the ship and there was not a moment to lose. One big sea would have finished us.

The ship was being run off before the wind, with the captain himself at the helm. The second mate was lying flat on the deck to wind'ard with the water swirling over him, hanging doggedly on to the tarpaulin to keep the loose hatches in place. Chips, naked except for his shirt, was sitting astride a corner hatch smiting furiously at the wedges with a hammer; while the watch were hastily endeavouring to pass a line from side to side to prevent the hatches being washed away.

The danger was imminent. If once the hatches had got adrift the seas would have poured down the hold and sunk us in a few minutes. We didn't need telling what to do; we flung ourselves on the hatches and, despite the green water pouring in on us, held them.

96

down in place while two or three of the men hastily knocked in a few wedges.

Well battered and half-drowned as we were, we dare not let go. Old Jamieson was swept off his feet and hurled against the crab-winch. He got a nasty gash on the head that stunned him, and Beckett and the steward carried him along to the foc'sle. The rest of us got the wedges driven home. The immediate danger was over, and we started to secure the hatch more thoroughly.

An old mooring-line was brought along from for'ard under the mate's orders. With this we took a number of turns round and round the hatch-coamings, sweating each turn tight with the 'handy billy' and frapping the whole lot together. This served to protect the wedges, but it was a long job, waist-deep in the pouring seas.

Eight bells came and went; the galley-fire was out; there was no coffee for anybody and the work went on. The whole hundred and twenty fathoms of line were served round the coamings, each turn frapped to the next, and finally lashed with lengths of rattlin-line tightly across the top of the hatch. It was near mid-day before the job was finished and the order given for the watch to go below.

All that day and night the storm lasted. Throughout the whole time the old man scarcely left the deck. Occasionally he went below to take a look at the barometer, or shelter himself for a few moments in the companion to drink a cup of coffee brought him by the steward. For the rest of the time he stood near the weather rail and watched the sea and the ship. He hardly spoke and, when he did, it was simply to give an order to the officer on watch.

Our masterful old skipper, I always thought, appeared at his best in heavy weather. Personally I always felt more comfortable when he was on the poop, and it seemed to me the barque herself behaved better under the direct eye of her commander.

He handled her wonderfully. Often, looking at our tiny sea-swept hull, with its towering fabric of spar and cordage, I marvelled how a man could drive that contrivance of steel and string across five thousand miles of stormy ocean and find his way to a particular cluster of human habitations on the other side of the globe.

A seaman, and a master one, was Captain West. Skilled in sea-lore and limbed like a Viking, he was no unworthy descendant of the Devon sea-dogs of old. His powers of endurance were extraordinary. When necessity arose he seemed able to do without food or sleep indefinitely. Grim and grey of face, with drenched side-blown whiskers and indomitable eyes, he would stand for hour after hour through day and night by weather-rail or helmsman, sole master of his ship and crew and, taking counsel of none, beat the storm-fiend at his own game.

On this occasion he kept the deck for thirty-six hours. It was the morning of the second day before the weather moderated sufficiently to enable us to make sail, and he went below and left the deck to the mate.

The maintops'l was the first to be put on her again and glad we were to be able to set it. We took the

halyards to the capstan and splashed round the wet
decks, sweating the leaches taut to the rousing tune of
'Reuben Ranzo.'

'Oh, Ranzo was no sa-a-ailor,
Ranzo, bo-o-oys, Ranzo!
They shipped him aboard of a whaler,
Ranzo, boys, Ranzo!'

With the lessening wind the barque began to roll
heavily, heeling over in sickening lurches, and we
hoisted the mizzen and maintopmast stays'ls to try and
steady her. The halyards of the latter carried away as
we hauled it up the stay and the block came down full
on Mac's head, hitting him an awful crack. It knocked
him flat but he was up again in a moment and jumped
to the downhaul, hauling away cheery as ever with the
blood trickling down his face and dripping off the end
of his yellow beard. Coming on top of what we had
just been through during the last forty-eight hours, with
nothing more than sea-biscuits to eat, I thought it a fine
exhibition of pluck and great-heartedness. Even Paddy,
who wasn't given to fulsome flattery, admitted Mac
had plenty of 'guts.' Some, as the poet says, are 'born
great' and even then can't live up to it. Mac was merely
an A.B., but he filled that far-from-starry sphere with
a right royal sufficiency.

This was the worst spell of bad weather we had
during the passage. We did not sight the misty tops of
Kerguelen, but edged a little to the north'ard and left

that lonely sentinel of the Antarctic well to the south. Snow and blow of course still followed us, but every day saw us farther to the east and we began to count the days that should bring us to our destination.

We boys were keenly looking forward to our first foreign port. We began to feel we had spent all our lives at sea. At least, we had learnt a lot since we took our last look at old England. The roaring forties are a fine training-ground for sailormen; not even the North Atlantic can beat them in the matter of gales and big seas.

The Third Mate, as I think I have mentioned, was a perfect encyclopædia of facts about the old-time clippers and their wonderful runs, and we listened to his yarns with avidity. The old man, too, in moments of expansiveness, could relate personal experiences of fast passages and critical happenings. He had been in command of ships since he was twenty-two years of age and for nearly a score of years previous to taking over the *Arethusa* had commanded one of the cracks of the Shaw Savill Line to New Zealand.

Unfortunately for the peace of the ship the old man and Paddy hated one another. Hardly a week passed but the stern masterfulness of the one and the furious bad-temper of the other produced a most terrific row. Paddy, I must say, never spared himself – or others. He was really a good seaman and he fairly boiled over when the skipper found fault with the stow of a sail or the trim of a yard and talked to him as though he were a clumsy hobo working his passage.

The culminating row of all occurred a few days after we passed the Crozets. A little thing started it. We had been hauling taut the braces and the old man found fault with the trim and told Paddy he was 'no seaman.' Paddy flared up and said something about the old man coming and trimming the yards himself.

'What's that you say?' shouted the old man.

Paddy repeated it: 'You'd better come and trim 'em yourself.'

The old man exploded at half-cock and the two went at it ding-dong in reverberating concussions of language, to the intense appreciation of the men. It ended by the skipper telling Paddy he 'wouldn't carry him as ballast' and ordering him to his cabin, where he went sullenly, with the thunderous threat hurled after him of being 'covered with broad arrows' as soon as we reached port.

The matter was adjusted between them somehow, for Paddy was permitted to resume duty next morning. The thunder-cloud still lingered, however, and we were on the verge of an electrical discharge every time an order was given and obeyed.

Whether the old man's temper had been frayed by his recent encounter with Paddy or not I don't know, but he was in a very truculent mood for weeks. In the longitude of Cape Leeuwin we hauled up a couple of points to the north'ard and shaped a course for Kangaroo Island. We passed the Cape, the most sou'-westerly point of Australia, on April 7th and a day or two later, when the air had grown considerably warmer and we were bowling comfortably along with the main-royal set, some genius discovered that the next day would be Good Friday. The men decided to ask the old man for a holiday, not, I fear, from religious motives, but as an opportunity to wash and mend their dilapidated clothes. Accordingly, at eight bells next morning they all marched aft, we boys dodging unobtrusively in the rear. The old man was majestically pacing the poop and Stedman stepped forward as spokesman:

'Beggin' y'r pardon, sir, but seein' as it's Good Friday,' he said, 'we're askin' you for a day's holiday, to do a bit o' washin'.'

The old man stopped in his walk, came for'ard to the poop rail and surveyed the men comprehensively up and down and all over.

'Do you see that truck?' said he, pointing to the golden ball on the main-masthead.

'Yes, sir,' said Stedman.

'Well,' said the old man, 'if the Devil himself was sitting up there, piping a general holiday, you'd still work. Now,' – and his voice rose – 'get for'ard, the whole damn lot of you!'

He resumed his stately walk and the men, realizing it was hopeless to look for a soft spot in him, marched for'ard again and were promptly turned to at scrubbing paint-work.

This was the only service I can call to mind ever held in the *Arethusa*. We growled, of course, and said hard things about the skipper, but 'growl you may, go you must' held good then and always, and there the matter always ended.

THE ROARING FORTIES

We made pretty good time across the fringe of the
Great Australian Bight till we were within three hundred miles of Kangaroo Island. Then the wind hauled
ahead and blew a dead muzzler. For eight long days it
hung in the east and constrained us to tack every two
hours, beating every inch of the remaining way across
the Bight. Coming at the end of a long passage it disgusted everybody. A solitary whale came up and
spouted near by, remaining in sight a considerable time,
while we hurled anathemas at it as the cause of all our
trouble.

The constant re-iteration of the order 'All hands
'bout ship!' made us very expert at the operation. Everyone knew his station and jumped for it smartly when
the word was given. Port watch for'ard, starboard aft,
and cook to the fore-sheet was the way of it. Then
'Ready-ho!' from the old man and down went the helm.
'Hard-a-lee!' and the fore and head sheets were let fly.
Up into the wind's eye came the ship and as she did so
'Mainsail haul!' and the great yards swung round, the
men taking in the slack and shouting for all they were
worth. As soon as the main was braced up 'Fore bowline!' was the order and the fore yards were hauled
round to the new tack. The ship gathered way through
the water, the yards were trimmed, the fore and main
sheets hauled aft and the gear coiled up. Twenty
minutes sufficed for the job and two hours later the
performance had all to be gone through again.

After about a week of tack and tack we sighted the
low shores of Kangaroo Island right ahead late one

afternoon. It was our ninety-seventh day out, and delighted we were to see brown earth after so long of endless salt water. As though it realized we had won through, the wind came fair and we ran in just close enough to distinguish the low white cliffs of the island and make sure of our position, then stood away on the opposite tack to round Cape Borda, the headland at its western extremity.

We passed the latter during the night, ran up the Investigator Strait and, when morning broke, found ourselves well into the Saint Vincent Gulf and in sight of a small island crowned by a lighthouse and signal station. We spoke the latter, reporting our name and port of departure, and at three in the afternoon sighted a low sandy shore right ahead, with a signal-station and a few scattered houses in the background. It was Port Adelaide.

Sail was shortened a few minutes later, the anchors put over the side and the ensign run up to the gaff-end. As we closed with the land we saw a steam-launch coming off with the customs officers and port doctor in the stern. A ladder was flung over the side and they clambered aboard. Down below they went with the old man, but were up again a minute or two later; the ship was rounded to, the tops'ls lowered, and the order given 'Let go!' Chips, standing on the cathead, knocked out the pin of the ring-stopper, the cable roared out and we brought up fast to our starboard bower in about twelve fathoms of water, distant a couple of miles from the land.

THE ROARING FORTIES

All hands were sent aloft to furl the sails. A 'harbour stow' we gave them, rolling the canvas into a neat skin as though it were covered with a jacket, and passing the gaskets at regular intervals like seizings. Then, after coiling down, all hands were knocked off for the rest of the day.

How strange the ship looked and felt as we came out on deck after tea for a smoke! After so many days with tiers of bellying canvas above our heads — to see nothing but the naked spars: instead of the unending dip and roll of the hull — to tread a motionless and steady deck: in place of the rustle of wind aloft and murmur of water overside — an uncanny silence. It all seemed very flat.

All night in was an unmixed pleasure though. We stood 'anchor watch,' that is to say, each one of us had an hour on duty and turned in for the rest of the night. My spell was from nine to ten, and after walking the deck and trying to realize that here we were — anchored on the other side of the world — I was relieved by Stedman and went to my bunk.

I slept like a top and effectually disproved the truth of the ancient jest, that when a deep-water sailor gets to port he is so accustomed to watch and watch that he has to get somebody to throw a bucket of water against the window and jangle a bell in his ear every four hours, or he would never sleep. It's a myth, he doesn't even dream.

CHAPTER SIX
Colonial Days
*

EARLY next morning a tug came off, passed her hawser aboard and towed us up to Port Adelaide.

My first feeling on reaching port was one of disappointment. Not at Port Adelaide, which was a bright, pleasant little place, with inhabitants hospitality itself, but from the very fact that we were in harbour again and that the sight and movement of the sea were shut out for a season. After all, the way of life in Adelaide was very similar to that which we had set out from, 14,000 miles away. The regular hours for work and meals, the traffic in the streets, the shops, even the very look of the buildings and appearance of the land itself, were not so very different. Australia was but another England, and orderly rows of houses seemed a poor substitute for long-ridged rollers, and a milkman's cart for a sea-shouldering whale.

Gone was the sense of vastness, of loneliness, of titanic grandeur or untroubled beauty, and all the magic and the mystery of the sea. In its place was the well-known routine of daily work and nightly sleep, of dusty solid earth and unimaginative days. 'Who hath desired the sea? the sight of salt water unbounded,' esteeming 'her excellent loneliness' rather than the comfortable ways of well-trodden cities? I know not what ancestral longing stirred in my blood, nor what dim avatar possessed me, but certainly I desired it beyond all that foreign lands could offer. The feeling remained throughout my sea-life and made me never sorry to exchange even the quaint picturesqueness of Japan, the mysterious jungles of Malaya, or the snowy Andes of Ecuador for the fascination of the lonely sea.

Not that I didn't enjoy our stay in Port Adelaide. Far from it. The Colonies are a very pleasant part of the world in which to spend a week or two, the Colonials the nicest people imaginable, and we took advantage of both the one and the other. There could not be a more delightful port of call for voyagers across the Indian and Pacific Oceans than Australia, and as a port of call I appreciated it. But when all is said and done, the earth is a globe mostly covered with water and though, it is true, pieces of land are stuck here and there about the seas, they are merely in the proportion of one to five, and this earth of ours is evidently intended for sailormen and the adoption of a seafaring life.

Nevertheless in the scheme of things ports fill a place and at this – our first – we looked with interest.

Our days from dawn to dusk were occupied in work on board. We had first of all to unbend the sails and rig cargo gear in readiness for the stevedores' men. A gang came aboard on the second morning, hatches were taken off and the process of discharging commenced. Only about half our cargo we found was for Adelaide, the remainder was consigned to Newcastle, our next port of call. A heterogeneous assortment of goods was soon being hoisted out of the hold. Cases of whisky and foodstuffs, crates of machinery and hardware, bales of various kinds, drain-pipes, cement and corrugated iron, were swung ashore in a continuous stream.

Stevedores did all the work; the ship's company had no concern with the cargo. We were all employed in chipping and painting overside and, as the holds began to empty, in scraping and painting the stringers and 'tween-deck beams. From six in the morning till the same hour in the evening our day's work lasted, after which we were free to do as we liked.

Most evenings we went ashore. We found Port Adelaide a cheery, go-ahead little town, very pleased to greet apprentices. The first place we made for was the Missions to Seamen, the padre of which had been down on board as soon as we made fast. It proved to be a snug little rendezvous, with a reading-room where writing materials were provided for all comers. Good use we made of the latter during our first few evenings in port. The Missions to Seamen, as I then for the first time realized, does a fine work for sailors in the seaports of the world. North, south, east and west flutters the 'Flying Angel,' assuring seafarers of a hearty welcome and all those English virtues and characteristics one misses after long sojourn in far seas and foreign lands. Its padres are the finest type of sky-pilot – real 'men' as well as parsons. Sailors as a rule, don't take kindly to clergymen, but the Mission padres are in a class apart, and so regarded. There are few seafarers who haven't at some time or other received a good turn from one of them. In the *Arethusa* we had to thank them for many a pleasant day and jovial evening from

Cape Town and Calcutta to 'Frisco and the States, besides other kindnesses of no less worth.

After the Mission the next place we hunted up was a restaurant, and were agreeably surprised to find we could get a good dinner of meat and two vegetables for fourpence. We didn't neglect that side of the amenities of shore-life, either.

The people proved very kind and friendly. They are much more free and easy than folk in England and seem to have a soft spot in their hearts for sailors. All sorts of people came down to look over the ship and a number of invitations to houses ashore were extended to us in consequence. There was only one other sailing-ship in port besides ourselves — the *Torrens*, a regular South Australian trader — so hospitality was concentrated upon us two. We went to several picnics and almost every other evening one of us was out to supper and a sing-song somewhere.

Two or three times we went up to Adelaide, the capital of South Australia, and a very fine city. The main thoroughfare — Rundle Street — is wide and tree-lined, with handsome public buildings standing in their own grounds on either side. The Torrens Gardens, the city's place of recreation, is a delightful spot with lakes and meadows and woodlands well laid out. It was moonlight the first time we went there, and a beautiful prospect it presented in the soft silvery light.

One Sunday Paddy and I set off early on a long excursion. Keeping parallel to the coast we had a twenty-five mile tramp in the direction of Port Wakefield. The coast was low and swampy, but a line of hills – the Gawler Range – was visible all the time inland.

We took our revolvers with us in the hope of getting a little sport. We didn't bag much; only, in fact, a couple of small birds like sparrows. Paddy possessed a little derringer and I, contrary to all regulations, had brought away a revolver and box of cartridges from which we had extracted the bullets and filled them up with buck-shot. The change was very detrimental to the rifling; I don't think it made much difference to the birds.

The last Sunday we spent in Adelaide it blew a gale. Several lighters broke adrift from their moorings, the Lightship in the bay parted her cable and drove ashore and the mailboat was delayed some hours by the short nasty sea that set up. We ran out extra moorings, made all snug aloft and congratulated ourselves we weren't at sea. The lightning was incessant and a tree close at hand was struck, but the ship sustained no damage. It was a new experience to be on board in a storm and to hear no cry of 'All hands!' no crash of breaking seas, nor be called upon to spend hours in a hard tussle aloft.

The gale had blown itself out by next morning and the work of unloading was not interrupted. We moved the ship over to Government Wharf during the day, as much of our cargo consisted of stores for the dock-yard there, and made fast under the stern of H.M.S. *Protector*, the new Australian cruiser that had just been added to the Imperial Navy. She was a very smart little vessel and the people were very proud of her.

COLONIAL DAYS

Several of our men cleared out that night. Amongst them were Johnson and the sailmaker, both of whom we could very well spare; Lopez also and Schmidt disappeared. Seamen were plentiful in the port just then, so the old man didn't trouble to search for the absentees but wrote them off as deserters.

The old *Torrens* was lying near us and several times we went on board her. She was quite a famous old craft and for long years one of the most popular passenger ships in the Australian trade. Her popularity was due to her fine sea-qualities, for she was not only very fast and dry in heavy weather, but could wing her way in light airs when other ships had difficulty in keeping steerage way. The famous novelist, Joseph Conrad, was for some time mate of her. She left port before us and had a very stormy passage home, ending up by running a vessel down in the mouth of the Thames – through no fault of her own. It was her last voyage under the British flag, for on reaching London she was sold to the Italians, to spend her old age under an alien name – the fate of many a fine clipper.

In three weeks our consignment of cargo for Adelaide was discharged, and with the last load swung ashore we started to bend sail again. The same evening we said good-bye to all the friends we had made, and quite a number of them came down next day to see us off. Our moorings had been singled up in the forenoon, and at mid-day the captain came on board with the ship's papers and four new hands in place of those who had deserted.

A little 'puffing Billy' of a tug was made fast for'ard, our mooring-lines were let go and as they fell with a splash into the dock, good-byes were shouted, the ensign was dipped and we were off.

An hour later we passed the Snapper, rounded Pelican Point, and commenced to set sail. Off Glenelg the tug let go and one after another the kites were piled on her.

For all our pleasant memories of Adelaide I jumped aloft with delight to cast loose the gaskets. Very different the setting out seemed from our previous one: that was only four short months ago, but we had learned a lot in the meantime. We boys looked on the royals now as our own property and could furl them and be back on deck ten minutes after the order to lay aloft; a downright disgrace it would have been for one of the men to have got there before us.

The food, too, was no longer nauseating, nor watch and watch a hardship; our stock of knowledge was increasing daily and we took pride in jumping to an order. Moreover, we stood our 'wheels' and 'lookouts,' and were learning to regard sea-boots as effeminate and to affect a 'hard-case' bearing in face of the day's duties.

A fresh breeze was blowing down the Gulf and we slipped gaily through the water, hoisting t'gans'ls to the tune of 'Whisky Johnny' and sending the royal yards aloft to a rattling hand-over-hander.

Glenelg soon dropped astern and with the breeze fair on the quarter and all sail set and drawing, a course was shaped for the inside passage and the watch ordered below.

It was fine to be at sea again, with canvas a-stretch and the lipper of water overside. The hull lifted, the good sticks leaned and I realized then what later I grew heartily convinced of, that it's a grand thing to have a hand to the wheel, the tops'ls sheeted home, and the world, the flesh and the devil fading out astern.

The wind being fair, the old man determined to run through Backstairs Passage, the strait between the eastern end of Kangaroo Island and the mainland. We managed to scrape through that narrow waterway successfully and a good many miles it saved us. All that night we ran steadily to the south-east, the coast of Victoria hull down to port and a spanking breeze a-rustle overhead.

We were kept at it till late clearing up the decks and getting rid of the inevitable litter of a spell in port. The old man was steadily pacing the poop; we had not seen much of him in Adelaide, now he let us know that he was still in command.

After glancing at us once or twice he came down the poop ladder and looked Gilroy up and down as though he found him profoundly interesting. It was disconcerting, particularly as the shirt 'Crackers' was wearing was none of the cleanest. The old man looked hard at it for a few moments, then told him to get a bucket of water and a broom. Gilroy did so and put them down in silence.

'Now,' said the old man, 'take your shirt and trousers off and scrub them. And see you get 'em CLEAN!'

He went back to the poop while Gilroy, who was

wearing nothing beyond the afore-mentioned shirt and trousers, stripped and spent half-an-hour on his knees scrubbing hard. He squirmed a good deal during the process, for the mates didn't trouble to hide their amusement and Gilroy found it difficult to be naked and dignified as well. He was feeling very aggrieved when later he trooped along to the half-deck, mother-naked, and with his wet clothes on his arm. Our light-hearted condolements didn't do much to soothe his ruffled feelings.

The breeze held fair and enabled us to make the passage of the Bass Straits and so avoid rounding Tasmania, which would have added some hundreds of miles to the length of the passage. At sunset on the fifth day out we sighted King Island ahead. It lies full in the fairway of the strait and has been the scene of innumerable wrecks. At midnight, we passed through the Narrows, off Flinders Island: the rocky cliffs come down on either side within a few cables' lengths of the ship. Fortunately the breeze held fair; to have tacked through such a place would have been well-nigh impossible. All hands were on deck for a couple of hours while the navigation was most intricate, but before dawn the last outlying rocks were well astern and we were stemming the waters of the wide Pacific.

At the first glance that ocean justified its name. It was only a glance though, for that very night it took pains to undeceive us and we had an astonishing electric storm. The sea did not rise very much but flashed and flickered all over with phosphorescent light. The

114

water sparkled so brightly that we hauled some of it up in a bucket to investigate the reason of its strange illumination, but were no wiser when we had done so. The weird shapes of corposants alit on our spars, uncannily running up and down in pale livid flames. The lightning display, too, was most curious — broad sheets erratically switching on and off like a gigantic electric light.

For an hour or two the disturbance lasted and proved the precursor of a stiff blow from the north-west, of the type known as a 'brickfielder.' The wind sprang up with extraordinary rapidity and the incessant lightning was no longer sheet but forked, while the thunder gave off a continuous ear-splitting roll of sound. To work aloft under such conditions was anything but pleasant, though a ship offers so many points to the electric fluid that she is seldom severely struck.

We made land on the port bow on the forenoon of the next day, the 23rd of May, and were off Newcastle Heads by eight bells in the afternoon. As we drew near we could make out the position of the port by the veritable forest of masts that rose up over the low fore-shore. Our tops'l and t'gallant halyards were let run as soon as we observed a tug coming out to us. There was a few minutes haggling on the part of the tug-boat skipper and our old man, but it didn't do to dawdle, for the breeze had hauled into the south-east and was blowing freshly, dead on shore. A bargain was quickly struck, the tug's hawser was passed aboard and, as she forged ahead, we laid aloft to hand our canvas.

Pretty soon it became evident we were not going to have things all our own way. Wind and sea were setting us down to leeward of the harbour mouth and, for all our footy little tug could do, we were unable to hold our own. The tug-boat struggled gamely and nearly burst her boiler in the attempt to hold us. But from the first it was a losing fight. We were showing a high side out of water, owing to the quantity of cargo we had discharged in Adelaide, and every minute the rising wind and sea set us nearer to the rocks on the northern shore.

The old man hung on till he saw there was no chance to weather them, then gave the order to loose tops'ls and t'gallants'ls and cast off the tug-boat. Damning the latter heartily, we let the hawser go, made sail and began to claw our way off the coast. To be so near and yet so far was exasperating, and the hopes we had entertained of a Saturday night ashore vanished with the receding shore-line. But worse was to come.

It was 'all hands on deck' — it had been ever since breakfast — and throughout that night there was no sleep for anybody. We spent the whole time beating off and on, going about every two hours. Wearisome work it was; we would beat up to wind'ard and, having got weather-board enough, stand in for the land again. It was after noon the next day — Sunday, to boot — before we were back in our old position off the harbour-mouth, with signals flying for a tug — a real one this time, and not a confounded kerosene-tin fitted with a screw.

We were so tired we could have dropped off to sleep

as we hauled; everybody had been on deck for over thirty hours, going hard all the time. The sight of a tug — a big, powerful boat — coming out, woke us up and we speedily got her hawser aboard and headed in for the harbour-mouth. Lying almost in the fairway, what should we see but the masts of a barquentine sticking up above water, which we were told had hit the rocks and foundered the previous night. We had had a gruelling, but fared better than she did.

It was growing dusk before we made fast. We tied up some distance up stream alongside the *Iredale,* the very ship that, strange to say, we had laid alongside in the East India docks five months previously. We didn't spend much time looking about us; as soon as the lines were fast, we rolled into our bunks, hardly stopping to drink a pannikin of tea, and slept the sleep of the just till five-thirty next morning.

When we turned out we were in a better condition to take stock of our surroundings. It was a busy scene that greeted us. The harbour was a wonderful sight by reason of the great number of deep-sea sailing-ships then in port. There were no less than a hundred and sixteen of them when we arrived, not counting steamers or coasters, and a grand show they made. Right away from Queen's Wharf, just inside the Bluff, up past the Dyke they lay in an unbroken line as far as Waratah, or 'Siberia,' as it was called, from its remoteness to every where else. In the Dyke, where we were lying, the ships lay three deep and there was a double row of them over on the other side at Stockton. Masts and yards were packed as thick as bristles on a hedgehog. During the day there was as much activity afloat as ashore, in consequence of the tremendous number of steam-launches, ferry-steamers, chandlers' boats and ships' gigs dodging about among the shipping.

The first thing we did was to put the gig overside. After breakfast three of us and Paddy pulled the old man up to town and landed him at the Custom House steps. It was my first experience as an oarsman and not a happy one. We landed the skipper without mishap and were told to return to the ship. A gentle breeze was blowing in our favour and, the boat being fitted with a dipping lug, Paddy determined to sail. It was an unwise decision and landed us in a howling mess in no time.

All three of us boys knew as little about stepping a mast in a seaway as a cow does of handling a musket. Our inexperience caused Paddy to curse frantically. The sail was set somehow and we had got perhaps a quarter of a mile on our way when an oar fell overboard. Having once got the sail set Paddy was loath to lower it, and we stood after the oar. Pick the wretched thing up we couldn't. We sailed over it or passed it out of reach or missed it altogether – anything but what we wanted to do. Paddy's curses only made us blunder more.

Our efforts were interrupted by a tremendous hail above our heads and, looking up, we saw the bows of a large sailing-vessel right on top of us. Paddy stopped swearing, grabbed an oar and sculled with all his might. We just managed to escape, but only just – cleared her cutwater but got a glancing blow from the bows that knocked us all in a heap. To an accompaniment of threats and howls we bumped and drifted aft, while a line of heads dotting the big ship's bulwarks gave us hot and hasty advice.

After that we soberly set to and recovered our oar. Paddy was like a volcano on the edge of eruption. He made us pull all the way back to the ship, and pull our hardest, under the fire of his scorching remarks. It was our first experience of boating, but before we left Newcastle we had learnt something about handling oar and sail.

We only stayed at the Dyke a few days, then shifted down to Queen's Wharf to discharge our cargo. Queen's Wharf was the best berth in port and only a couple of minutes' walk from Hunter Street, Newcastle's principal thoroughfare. Just ahead of us lay the lofty *Andorinha*, her skysail yards almost out of sight, and astern the *Forfarshire*, one of the cracks of the Shire Line.

Stevedores again did all the work of unloading, the crew being employed in chipping and painting overside. We boys were put cleaning brasswork and attending to the gig.

A couple of mornings later I felt very sick when I woke and didn't turn out with the others. The old man sent for me in the course of the morning and gave me a tumblerful of castor oil. He informed me as he did so that he wasn't going to have any sick boys on board and, if I didn't turn to, I could pack my bag, and go to hell and the hospital. I turned to: and decided next time I laid up it should be something fatal, lesser ailments weren't worth it.

Castor-oil – how we hated it! The beastly stuff was Captain West's unfailing remedy; he applied it for every complaint human flesh is heir to. On one occasion Nils fell down the hold; the poor beggar fell a full thirty feet and dropped across the keelson. We hauled him up on a hatch and by the time we got him on deck the old man was along with a stiff dose of castor-oil. By good luck Nils had no bones broken, and, after two or three days in his bunk, was able to get about again. I have no doubt the old man considered it was his castor-oil that did it: he would have used it for resuscitating a corpse.

Most of our evenings we spent ashore. We found Newcastle a very lively and pleasant little town; with so many ships in harbour the atmosphere of the place was of the sea salty. Badge-caps of apprentices were thicker in Hunter Street than blackberries on a hedge in September. Our own houseflag and badge was a diagonal silver bar on a red field and, the ship not being notorious for high living, was known as 'the white streak of starvation on the Red Sea of misery.' Besides apprentices there were any number of officers and seamen about too, but the reefers, being in uniform, made the gallantest show. It would have taken a bold man to shout 'Light the binnacle, boy!' – the epithet often called after a brassbounder in sailor town – in Hunter Street on a Saturday night.

Different cliques and different companies had their

favourite haunts, which strangers didn't enter uninvited. There was one hotel, the 'Clarendon,' which was common property. It was the best-known hostelry in town, chiefly in consequence of the popularity of a barmaid there – Nell, by name – who was often known to present half a sovereign to a hard-up customer. Her fame at that time spread over the seven seas, and discredited indeed was the man who hadn't drunk a schooner of beer at the 'Clarendon.'

A good meal at a restaurant, a walk round the town, the making of a few purchases from the scanty funds doled out to us by the skipper, and a glass of beer at 'The Pilgrim's Rest,' made up our evening's entertainment. There was rarely anything more riotous than high spirits and practical jokes – for one thing everybody was always hard up. The Colonial beer, though, had a reputation for headiness. 'Sheoke' beer it was called and old Australian traders used to spread a net under the gangway, called therefrom the 'sheoke net,' whose office it was to save mariners who 'missed stays' when coming aboard from falling into the dock. Each morning the ends of the net were hauled up and its slumbering occupants bundled out on deck.

A great institution was the sheoke-net and saved from a watery end many a mariner whom it would be harsh to call drunk. 'Drunk' is an opprobrious term and implies a state of intoxication absolute and irredeemable. What it means to be really 'drunk' is well-suggested by a little yarn that often went the rounds.

It is said that late one night in Melbourne a well-primed seaman returned on board his ship, staggered up the gangway and collapsed on deck, where he lay sleeping peacefully. The mate came on board soon after and, seeing somebody 'dead-oh,' whistled for the night-watchman. 'What the hell's this?' said he, 'take this man for'ard; he's dead drunk!' The watchman bent over the prostrate form for a moment, then looked up cheerfully, 'No, sir, 'e ain't drunk,' he said, 'I saw 'im move!'

The Third Mate left us before we had been long in Newcastle. He was paid off by the old man, and personally, I saw him go with more than a tinge of regret. He was a good sailor, bold and very daring, but his temper was like a wildcat's and the atmosphere was certainly clearer after his departure. He stayed ashore for a few weeks, and he and I had some long tramps together into the country and up the Hunter River. Then he signed away as Third Mate on a hard-case Yankee barque. The unofficial designation of the Third Officer on those ships was 'blower and striker' and Paddy would have filled the post to a nicety. His ship sailed before we did and I never saw him again.

About this time we had a succession of hard gales of wind. From our berth at Queen's Wharf we could see the harbour-mouth and the white-topped racing seas outside and it was matter for congratulation to us that we were snug in port. Other ships were not so fortunate and there were quite a number of wrecks and mishaps on the coast about that time. A steamer went ashore just south of the Bluff and a schooner was dismasted, but the most spectacular happening was the attempt of the *Norma*, a four-masted bald-headed barque, to enter port under her own sail. She ran for the harbour without waiting for a tug, but failed to weather the reef on the northern side of the entrance (the still-standing sticks of the barquentine might have warned her), and was within an ace of coming to grief. She made a running moor to seaward of the rocks, let go both anchors and brought up within a stone's throw of disaster. She was in the backwash of the broken water from the reefs and we looked every minute to see her cables part and herself drive ashore.

The lifeboat went out and made fast alongside, but the skipper wouldn't abandon her and signalled for a tug. Several went to her assistance but couldn't do anything, hawser after hawser parting like rotten thread.

The cables held and the lifeboat stuck to her. All night she lay almost in the smother of the reef, and next morning the *Champion*, the most powerful tug on the coast, arrived from Sydney. The latter got a $4\frac{1}{2}$-inch steel wire hawser aboard and after a hard tussle hauled her out, made a wide sweep in the offing and towed her safely into port. For the twenty-four hours she had been in her perilous position she was in full view of a large part of the shipping in port, and we cheered her and the *Champion* lustily as she towed past to her berth up the Dyke.

Soon after this, our cargo being all discharged and freights so low there was little likelihood of a charter, we took in some ballast and shifted ship to a lying-up berth

at Siberia. We towed up past all the shipping, and a
brave show they made. Every sort of rig and nationality
was represented. There were tall and stately British
clippers, four and five-masted American schooners,
white-painted South Sea traders from the Islands, huge
port-painted Frenchmen, sharp, black-hulled Germans,
hard-case Nova Scotiamen, Italians, Norwegians and
Danes — graceful clippers and lumbering sea-waggons,
barques, barquentines and schooners, lofty or squat,
smart or lubberly, with a sprinkling of steamers and
coasting-vessels and a whole medley of smaller craft.

Arrived at Siberia, there began for us a long wait.
Our days were occupied in the interminable job of
scraping and painting the ship inside and out. The men
were at it all day, but we boys spent a lot of our time in
the boat. We pulled the old man up to Newcastle every
morning and brought him back for dinner. Often we
were out in the afternoon as well on fishing excursions.
We pulled leisurely up the Hunter River amid very
pretty scenery, stopping every now and then to fish.
We made some good hauls, mostly of small kinds of
which I never learnt the name. It was a much more
pleasant way of spending the afternoon than on a stage
slung over the ship's side, tapping away with a chipping-
hammer.

We were so far out at Waratah that we only went up
to town on Saturday nights. It was a three and a half
mile walk and no easy one at that. The first part of the
way was over wide swamps, half-filled with rubbish and
a perfect paradise for mosquitoes. Thereafter we had to

walk the whole length of the Dyke, with its endless line
of ships; it meant jumping over mooring-ropes and
wires every few feet and was more like a hurdle-race
than a walk. Finally we took a ferry-boat and were
landed at the foot of Hunter Street. And when the
evening was over we had the whole steeplechase back
again.

Sundays were our best days. There was no work
doing then and lots of people came down on board,
after the pleasant manner of the Colonies, to look over
the ship. They came in regular parties and we had
great times showing them round. Invariably they asked
us to come up and visit them at their homes and pleasant
evenings the return calls always resulted in.

After nearly six weeks at 'Siberia' we heard one
morning that we were chartered, and proceeded to shift
ship down to the coal-tips to take in our 'stiffening' —
just sufficient coal, that is, to ballast the ship. It only
took a few hours to shoot aboard and then commenced
another wait, this time lying moored out in the stream,
to discharge our stone ballast.

The ballast was in the form of broken bricks and
rubble and proved hard work to discharge. It had to
be shovelled into baskets, heaved up on the winch and
emptied into lighters alongside. Having fairly fine lines,
the *Arethusa* needed a deal of stiffening and much care
to keep her in trim.

Lying as we were in midstream we were not able to
go ashore, but visited some of the ships near us and got
very friendly with their apprentices, especially on the

Inverkip. There were four of the latter, all wild Scotsmen and very nice fellows. We spent alternate evenings in each other's half-decks. Their skipper was a good sort and had his wife and children with him; altogether the *Inverkip* had the name of a happy ship. It came as a grievous shock to us when we arrived home to learn that she had been run down by a steamer, only a week or so previously, when homeward-bound, some sixty miles south-west of the Fastnet. Everybody on board was lost except the carpenter, who managed to scramble up the steamer's bows as she backed out. The news grieved us heartily; 'lost with all hands' is a bitter epitaph for a ship, and the frequency of it lends a grim propriety to the Mariners' Society's motto: 'There is sorrow on the sea.'

Ships were daily leaving Newcastle, bound, for the most part, to the West Coast, and on the 10th of July, after nearly eight weeks in port, we got our orders and heard we were chartered for Callao. The news pleasantly surprised us, for Callao is about the best port on the western seaboard of South America.

We found we had a foul hawse to clear when we came to heave the cables in. We had boxed the compass every flood and ebb and the result was a fine 'hurrah's nest' to disentangle. Four hours it took us, shackling and unshackling, riding to a creaky coir spring, and swinging in a bowline under the bows. 'Clearing hawse' is a job richly productive of accidents, but we managed without mishap and towed down to the coal-tips.

Loading, once started, didn't take long. The lines

were barely fast before the first truck-load was rattling down the shoots. For the next thirty-six hours we were hidden under a black pall of dust and all hands were kept busy down below, trimming the coal as it came in. Half-choked and black as sweeps, we were sufficiently thankful when the last load had been hurled below and we hauled off to the buoys to prepare for sea.

A wash-down worked wonders with the barque's appearance. Sail was bent, hatches battened down, ports and spars securely lashed and the boat hoisted aboard. The old man came off in the agent's launch in the evening with four new hands, those we had shipped in Adelaide having run directly on arrival. Two of the new hands were powerful and sailorly Norwegians, the third a Welshman and the fourth a young American who had just cleared out from the *Desdemona*, a fine old clipper lately arrived.

Next morning was fresh and breezy, with a cold snap in the air. We singled up our moorings and rode with a slip on the buoy, till a tug came off and was made fast for'ard. At mid-day order was given to cast off. A few minutes later we towed past the Bluff, sail was crowded on her and almost as soon as her forefoot lifted to the first long Pacific swell the tug was let go, tops'ls and t'gallants'ls sheeted home, and we were off once more.

I never felt anything more delightful than the first gentle rise and fall of the ship to the 'scend of the sea. After so many weeks of lifeless equilibrium, the slanting decks and long resistless heave and roll of the hull, a-dip to the wash and tremor of ten thousand miles of ocean,

gave one a sense of freedom and exhilaration impossible
to describe. As sail after sail was sheeted home and the
barque laid steeply over to the freshening breeze, I, for
one, would not have changed places with the skipper of
a battle-ship. Each to his taste, and mine for the open
sea and a square-rigger setting sail.

CHAPTER SEVEN
The South Pacific
*

THE breeze freshened as we drew away from the land and by nightfall we were plunging along under t'gal-lants'ls and the last trace of land had faded out of sight astern. Our course was shaped for the Cook Straits and when, at eight bells, we hove the log eleven knots were over the rail before 'Stop' was shouted.

Right across the Tasman Sea the breeze blew strong and fair; at times it rose to a fresh gale and we had to snug the barque down to upper top'sls. It never shifted from the quarter and we ran before it like a scudding cloud. On Wednesday the 22nd, only seven days out, we sighted the rocky shores of the southern island of New Zealand, and within a few hours of lifting their lofty ramparts the good breeze dwindled and then chopped right ahead.

Tack and tack we went at it; going about every four hours. Cook Strait lay right ahead and we could not afford to make an inch of leeway. During the night the wind veered aft again and when dawn broke the lofty mass of Mt. Egmont was full in sight, shimmering high up in the northern sky – a white, snow-shrouded cone. We stood in, and again the wind hauled ahead and we made Cape Farewell on the next tack.

Cape Farewell Captain Cook called it, for it was his point of departure. We didn't seem able to get away from it and when, on the other tack, we dropped it astern, there, right ahead, was old Egmont standing up like a gigantic beacon to the nor'ard. If a beautiful sight could compensate for a muzzling wind we were no losers, for Mt. Egmont is a grand peak, 8,000 feet high and so near the sea that it looks loftier still. Captain Cook estimated it as equal in height to Teneriffe; in reality it falls short of the latter by about 4,000 feet.

Three days we wasted beating back and fore, then, on the 25th, the wind hauled fair and next day we reached the Narrows between Wellington and Picton.

Wellington was little Tommy's native place and he fairly bubbled over with pride and excitement as the entrance of Port Nicholson opened out. We were not putting in and Tommy had to content himself with stopping all and sundry and pointing out the spot 'where I come from.'

We stretched over to the southern coast soon after, where high, bleak mountains run down close to the sea, with rollers breaking at their base and their summits hidden in cloud. It was dark before we passed the Brothers Lighthouse and headed out for the wide stretch of the Pacific. The moon rose later and to starboard we could plainly see the bluish, snow-clad peaks of the Kaikora Range, a hundred miles distant. Touched by the silvery shafts of light the ridge of mountain crests stood out like a glimpse of fairyland.

All that night we heard a curious scratching noise coming from the rocks to the south'ard of us; it sounded very weird and suggested the scraping of multitudes of tin pans. The old man said it was the calling of penguins.

Two days later, the last of July, we crossed the 180th meridian. As we were travelling from east to west this meant an extra day in our reckoning and we had two consecutive Wednesdays. We hoped we might have had two Sundays, for Sunday was a day of leisure, but no such luck. Stedman said he had never heard of a ship crossing the 180th meridian on a Sunday: he doubted if the thing were geographically possible. The fact of the matter, of course, is that twenty-four hours has to be sandwiched in somewhere and, within limits, it is entirely at the option of the commander what day he chooses to call it. Still, it never is a Sunday.

The Third Mate having left us and his office being vacant, as soon as we left Newcastle the captain raised Stedman to the status of bosun. In a windjammer the third mate is often described as a 'glorified bosun' and Stedman took over most of Paddy's duties, though he still continued to berth for'ard. There was a vast difference between the two men and the change made for comfort. Paddy was a regular mandriver; Stedman, an equally good sailor, had as a principle 'Duty is duty, and bullying isn't.' He was the best type of British seaman — self-respecting, a master of his work, and reliable as the sheet anchor. In an emergency he was as good as two men aloft, for he was always cool and never lost his head. When we were bending the mainsail in Adelaide the port lift, by which the yard is kept square, carried away. Most of the men were on that side but Stedman was at the starboard yardarm hauling out the earring. The moment the lift parted the ninety-foot yard cockbilled. The weight of the men swung the port yardarm down to the rail, while the starboard soared swiftly up. Stedman missed being crushed by the tops'l yards by a hair's-breadth and, though the yard was almost up and down, hung on by his legs and made fast the earring. If his nerves had not been so steady he never stood a better chance of being killed.

After getting through the Cook Straits we had a perfect orgy of catching albatross and molly-hawks. Fifty-four of the latter we hauled aboard in one day, all of which we boys were set to pluck, watch on deck and watch below, to complete the skipper's mattress.

Birds and fishes are very plentiful round New Zealand. Besides albatross, gulls, and petrels, there are great numbers of boobies and gannets. Wonderful divers gannets are. Their beaks are but the prolongation to a point of the lines of head and body. Flying high, a gannet will suddenly fold its wings close and drop vertically like an arrow from the sky. It is up again a moment later with a fish in its beak.

Cruising about among the other birds were usually several frigate or man o' war birds. They are beautiful fellows, but regular pirates in their habits. They look like freebooters, lean and rakish. Their tails are forked and their outstretched wings shaped like an elongated W. Immensely powerful of wing, they never alight on a vessel and apparently keep the air for weeks at a time. They neither swim nor dive, differing therein from all other sea-birds and, in consequence, have a very curious way of procuring their food. They watch the smaller birds as they fish and directly one of the latter has made a catch, swoop down on it. The luckless fisherman lets go its prey, but before the latter has time to reach the water, the frigate wheels like a flash of light and catches the fish in its beak.

Watching birds and fish, I always thought, is not the least interesting part of an ocean passage. Numberless species are to be seen from the deck of a leisurely wind-jammer; shy creatures that never approach the throbbing pulse of a propeller. One might make many voyages in steam and never see a turtle, or even a frigate-bird on the prowl for its dinner. None of them, whether fish or fowl, show any dread of a clipper's white wings; in all probability they take her for some gigantic sister of their own.

After leaving Cook Strait we steered about south-east till we were on the 45th parallel of latitude, when we altered course due east to run our long easting down. It was near mid-winter and the weather grew bitterly cold. For several nights in succession we saw the Aurora Australis, or 'Southern Lights,' as sailors call them. They are streamers of different coloured light shooting up from the horizon almost to the zenith. Soon after dark they appear, at first low down on the polar horizon. Concentric circles, like rainbows, rise one above the other, in bright bands of changing light, the outermost being sometimes surrounded by a fiery halo. Up and through these circles shoot bright flicker-

ing spears and shafts of light. The whole display
alternately dims and brightens. It may last for half an
hour or so and then the radiance seems to dissolve and
ebb gradually away. Sometimes the 'Lights' are so dim
as to be barely visible; at others they cover half the sky
with a splendour like the unfolded wings of seraphim.

Glorious displays of light and colour are to be seen in
these Australasian waters. We had had magnificent
sunsets coming round the coast and again when crossing
the Tasman Sea. There, too, for the first time I saw
the 'Green Flash,' as it is called. Just as the sun is
about to sink below the horizon a flash of vivid green
seems to leap from it. It only lasts a second and is gone.
What causes it I do not know, for often the sun sets
without it. The old man used to say it was 'the sun
putting out his sidelights,' the emerald green representing a sailing-ship's starboard light. Of course it is only
visible when the horizon is absolutely clear.

Beautiful as sunsets are, sunrises are even more so.
To my way of thinking, sunrise over a tropic sea is the
most glorious sight on God's earth. But it's too common
an occurrence – that's the fault of it. It happens every
day, whereas if it only occurred once in fifty years
travellers would come from all over the world to marvel
at it. To see in a few moments the velvety blackness of
night tremble through veils of paling purple, hueless
grey and all shades of azure, delicate rose and flashing
gold to the imperial blaze of the risen sun is an unforgettable experience. Sunrise anywhere is a glorious sight
but it puts on its most ineffable beauty, I think, at sea

between Cancer and Capricorn. Looking at it one feels the same sort of illumination as the prophet of old must have done when he exclaimed: 'God is light!'

As we drove farther to the south we approached realms where it is as rare, as it is pleasant, a thing to behold the sun. We passed Chatham Island on the 5th of August and soon after met with a succession of gales that chased us right across the breadth of the Pacific. We encountered a mighty old whale – a cachalot – near Chatham Island, that some of the men were ready to swear was responsible for the change.

The old fellow was heading due South and coming along leisurely, heaving his great bulk half out of the sea and seemingly intent on 'making a passage.' He was bound straight for the pole, and in the immensity of his determination and the apparent deep contentment of his 'blows' and plunges, gave one the impression that he found life a joyous adventure. His course was as true as a ship's; time apparently was no object to him; he was bound for the pole, and having made up his mind to it, didn't seem to care whether he got there this year, next year, or the year after.

Whether he was to blame for the change or not, a change we certainly had. Right 'roaring forties' weather came down on us with a swoop. I nearly missed experiencing it or anything else beneath the glimpses of the moon, for in the first blow that caught us I had a very narrow escape.

It happened one night. I had laid aloft to furl the main t'gallant-stays'l and, as our custom was, had thrown my leg across it to ride it down. Being pitch dark, the head downhaul had not been hauled properly taut and, before I could get a turn with a gasket, the head shot up the stay, the sail gave a flap, and I was knocked off.

Instinctively as I fell I threw up my hands and, by what was little short of a miracle, grabbed hold of the lower foot-rope and held it. The jerk almost pulled my arms out but I managed to hang on and, twisting my legs round the stays'l sheet, slid down to the fiferail and so stepped on deck.

It was about the closest shave one could have had. All that night my body seemed on fire, and though I was allowed to stay in my bunk, sleep I could not. Lucifer himself never fell so far as, all night long, I kept doing in sudden fits and starts. I turned out next morning, but the wrench had made my shoulders black and blue, and for a few days I was as weak as a kitten and kept on feeling horrible jerks at the back of my head.

Being winter time, there was not only the cold to put up with, but darkness as well and, of the two, the latter was the more trying. The wan and sickly daylight lasted a bare seven hours out of the twenty-four. For the rest of the time we were plunged in complete darkness, a cold, hail-smitten darkness, black as the Earl of Hell's riding-boots.

Day after day the sou'-westerly gale stormed at us and we surged through mountainous seas, with snow and hail-squalls beating down on us with damnable iteration.

We worked and ate and slept in our oilskins. Little use they were though, for all the clothes we possessed were soaked and some of the hardier spirits gave up the pretence of wearing oilskins at all. Life alternated between wet bunks and wetter decks, punctuated by icy spells aloft, with wearisome monotony.

The *Arethusa* was at her best in stormy weather and behaved nobly. At times a fiercer squall than usual would lay her over almost on her beam-ends, while the wind's note rose to a scream and the hissing, hail-whipped water swirled in over the lee bulwarks. Green seas like cliffs crashed over the weather-rail and made a maelstrom of the decks, forcing all hands to jump and hang on for their lives.

All work, save the handling of the ship itself, was at a standstill. The barque required all our attention. Buffeted and weary, the watch on deck were kept at it unceasingly attending to her needs. On deck and aloft, there was constantly something to do — lashing a spar or a boat, securing madly-swinging ports, passing extra rovings, or restowing a sail that threatened to break adrift.

The sea was a wonderful and magnificent sight. The rollers, high as our foreyard and stretching north and south as far as the eye could reach, were titanic in their grandeur. Under the lash of the hurricane squalls the seas turned to fields of snow just streaked with ribbons of dull green. There no longer seemed to be any waves, the whole surface of the sea assumed constantly-changing inclines and the barque's plunges grew longer and steadier through her very insignificance compared to those undulating mountains of water. It was impossible to estimate their height, but fifty feet would certainly be no exaggeration. As the barque slid down some foam-topped, league-long slope, the water ahead rose higher than the foreyard. Next moment her bows were pointing skyward and her stern fell into the trough, with a mighty wall of water rising up behind her. It is a nerve-shaking sight to look astern and see a hissing roller, as steep as the side of a house and as high, rushing towards the ship. Just when it seems it must fall with a cataclysmic crash and overwhelm her, her stern rises swiftly to it and the foaming crest rushes under her counter and surges for'ard, balancing the barque like a cork on its crest. Ships that carry dago crews often rig a canvas screen behind the wheel, so that the helmsman may not lose his nerve at the sight of a toppling, following sea, but the sterner seafarers of the north despise such weaknesses.

For three days we ran with only lower tops'ls and foresail set. Only our best men – Neilsen, Mac, Jamieson and the Norwegians – were allowed at the helm, with another of us to leeward to put the wheel a-weather or a-lee smartly. The wildly-swinging compass-card was useless and for sixteen hours the old man conned her by the sweep of the sea.

This storm blew itself out, but for four weeks the wind hardly ever dropped below the force of a fresh gale. Once, a sea broke aboard with such force that the spray was dashed as high as the lower top'sl yard – a full seventy feet – and drenched the clew of the upper-tops'l. On another occasion we were struck by a very heavy squall – the Pacific, in spite of its name, being a regular happy hunting-ground for squalls. This particular specimen shook us in its teeth, so to speak, and sprang our upper maintop'sl yard for us. It happened at four bells in the morning watch – 6 a.m. in other words. All hands were at once called and it was supper-time before the yard was properly secured and the watch sent below.

A handsome day's work it gave us. We rigged a gantline and swayed aloft, one at a time, our heavy iron awning stanchions. These were placed lengthways all round the yard over the damaged place and securely lashed with chain. The lashings were then tautened by means of oak wedges driven in with a topping-maul.

The operation is known as 'fishing'; it took us the whole of a bitter winter's day and added almost a ton to the weight of the yard. The big sea that was running and the way the barque tumbled about, made it an arduous undertaking and poor old Chips had his hand crushed while clearing a stanchion that fouled the lower tops'l yard. But it was finished soon after dark and the repair lasted until we reached home, and only needed to have the wedges tautened occasionally.

Hard, stormy weather followed us till near the ninetieth meridian, when we hauled up and steered more to the nor'ard. One dark, moonless night just before we got clear of the 'forties,' with a fresh breeze blowing and the ship running quietly along under t'gallants'ls, there occurred a most uncanny experience.

It was about four bells in the middle watch, the 'churchyard' watch, as the four hours after midnight is called, that it happened. We of the mate's watch were on deck – the men for'ard, Beckett and I under the break, and Mr. Thomas pacing the poop above our heads. Suddenly, apparently close aboard on the port hand, there came howling out of the darkness a most frightful, wailing cry, ghastly in its agony and intensity. Not of overpowering volume – a score of men shouting together could have raised as loud a hail – it was the indescribable calibre and agony of the shriek that almost froze the blood in our veins.

We rushed to the rail, the mate and the men too, and stared searchingly into the blackness to wind'ard. The starbowlines, who a moment before had been sleeping the sleep of tired men in their bunks below, rushed out on deck. Shipwreck would hardly bring foremast Jack out before he was called, but that cry roused him like the last summons. If ever men were 'horror-struck' we were.

Even the old man was awakened by it and came up on deck. Everyone was listening intensely, straining their eyes into the blackness that enveloped us.

A moment or two passed and then as we listened, wondering and silent, again that appalling scream rang out, rising to the point of almost unbearable torture and dying crazily away in broken whimperings.

No one did anything, or even spoke. We stood like

stones, simply staring into the mystery-laden gloom. How long we peered and listened, waiting for a repetition of the sound, I do not know. But minutes passed and still it did not come, and slowly, like men coming out of a trance, we began to move about and speak to each other again.

We heard it no more and gradually, one at a time, trickled back to foc'sle and half-deck. As far as the occupants of the latter were concerned, no one evinced any inclination to turn in and we sat around, smoking and discussing what the sound we had heard could possibly be. Nobody slept much more that night and thankful we were when the grey dawn broke over the tumbling, untenanted sea.

This was all. In bare words it doesn't sound very dreadful, but it made that night a night of terror. For long enough afterwards the echoes of that awful scream would ring in my ears, and even now it sends a shiver through me to think of it.

Who and what it was that caused it we never learnt. We hazarded a variety of guesses, many of them far-fetched enough. The cry of a whale was suggested, but I never heard a whale utter any sounds with its throat. Some other sea-monster, somebody else thought, that only rarely comes to the surface – but this was more unlikely still. The screams of seals or sea-lions on an island beach was another hypothesis – again, the nearest land was Easter Island, six hundred miles to the north-'ard. Besides, the shriek we heard had certainly a human, if not a diabolic origin. Whether it was, as some imagined, a shipwrecked boat's crew who saw our lights and in their extremity raised a sort of death-scream, or whether, as others asserted, it had a super-natural origin, remained a mystery insoluble.

The passing of this goblin-haunted night marked the end of the bad weather encountered on the passage. Thereafter, as we slanted north, there grew a daily improvement. The wind steadied, the sun shone for longer intervals and the seas no more poured aboard in green and solid cataracts. Royals were set, life-lines unrigged and our salt-soaked 'soul-and-body lashings' laid aside.

On the 20th of the month – thirty-seven days out – we had our first glimpse of civilization again in the shape of a large barque spooming along under full sail and homeward-bound. We envied her greatly, for with reasonable luck she ought to have been home for Christmas. She was steering straight for the Horn and coming along grandly, a white ridge of foam curving away from her sharp forefoot – 'carrying a bone in her teeth,' as sailors say. We did not speak her and in less than an hour she was out of sight.

She was the first vessel we had seen since leaving Newcastle. There was nothing unusual in that, for little troubled by the fleets of commerce are these wild heights of the mid-Pacific. Only a wandering wind-jammer like ourselves seeks their fastnesses; even New England whalers are few and far between; and no scrap of land breaks the mighty expanse of ocean between Macquarie Island and the Horn.

A few days after sighting the homeward-bounder, when the south-east trades had folded us in their welcome sweep, we raised the small islands of St. Ambrose and St. Felix. They lie in about 27° South, some six hundred miles off the South American coast, and gave the old man an opportunity of adjusting his chronometers.

From the look of them it will be long before they are used for any other purpose than that of correcting the time. Lying on a submarine ledge that forms the western boundary of the Peruvian Sea, and separated from the Continent by the abysmal trench of the Atacama Deep, they are mere platforms of rock, verdureless and, as far as we could see, uninhabited. The long Pacific rollers have worn their windward coasts into curious shapes. They are fretted into caves and grottoes, with overhanging arches and isolated pinnacles. The broken-off rocks make the weather side of the islands quite inaccessible. San Felix affords anchorage of a sort on its northern shore under the lee of a rounded hill and is, I have heard, a great place for lobsters, but there isn't a drop of fresh water in the whole extent of the island. Gonzalez is merely a bare and bird-haunted hummock and San Ambrosio is more inhospitable still. Its precipitous cliffs rise sheer out of the sea, with deep water right up to their sides. Sixteen hundred feet high, steep, sun-scorched and with a wicked reef at its eastern end, it belies its fair immortal name. It is quite inaccessible and the foot of man, I believe, has never once landed on its surface. Its humanity is as questionable as its divinity.

Besides these three larger islands there is a smaller mass of rock known as Peterborough Cathedral, from a fancied resemblance to the majestic west front of that building. The water-worn rocks do distantly remind one of Gothic arches and pillars and I wondered what homesick Fenman had first given rein to his imagination and boldly christened it Peterborough Cathedral.

With the coming of the south-east trades the old man resumed his fishing operations. He turned to fishing for recreation in the same way as most people pick up a book. In fine weather the captain of a sailing-ship has long hours on his hands. Some skippers take up carpentry or other useful craft, some read, some simply loaf and a few drink. Our old man fished. He fished for bonito off the bowsprit, for sharks and barracoutta over the stern; he lay in wait for porpoises with a harpoon on the foc'sle head, and perched himself on a bumpkin with the grains for any mortal thing that came within reach. He would 'fish' for sea-weed, frigate-birds or whalefeed and even haul a bucket of water up and examine that.

But with all the fishing that we shared in, we boys were not permitted to neglect our duty. As often as not the procedure of catching something would take place in our watch below. By the time the excitement was over it would be the other watch's turn below, and heaven help the apprentice on duty who was caught running into the half-deck! It meant staying on deck an hour after one's own watch had gone below and working with the other watch. It doesn't sound much of a punishment, but in practice it was devilishly unpleasant.

144

Old West saw to it that his sporting predilections did not interfere with the discipline of the ship; I never sailed with a man who ruled his crew with a firmer hand. No one was ever permitted to sit down on deck or to speak when at work. His occasional tours of inspection for'ard were disliked by the men and dreaded by us. If the condition of the half-deck did not satisfy his searching eye the watch below would be galvanized into activity, and turned to at once with buckets of water and sand and canvas to scrub for an hour or so. We never knew when he would pay us a visit, and had in consequence to keep our deck always a spotless white and the paintwork with never a fingermark.

So with the men. Any dilatoriness or sullenness on their part would inevitably result in a 'work-up' job. Sure as fate, all hands would be called and a couple of hours spent in the rapid performance of some highly unnecessary operation. Once, in consequence of an insubordinate demeanour on the part of the men, we shifted our whole suit of canvas aloft, twenty-one sails in all, during the hours of darkness. Five nights it took us, working six hours a night, with all hands on deck. For the rest of the time we were watch and watch. It nearly broke our hearts, but the old man saw to it there was no shirking and carried the day by sheer masterfulness.

An odd mixture of kindness and bullying was his treatment of us boys. He would work us till we were ready to drop and then, as likely as not, call one of us along and lend us a Findlay's Directory or some other nautical treatise to read. On this passage, running up

the Trades, he had the four of us along in the dog-
watch and taught us how to make a variety of knots and
different kinds of plaiting. We put complicated 'Turk's
heads' and 'Matthew Walkers' on the handles of the
buckets at the poop rail and made a variety of mats —
sword-mats, sennit mats and sheathing for the 'Scotchman' — to serve as chafing-gear aloft. It was mostly
done in our watch below; to have permitted such tasks
to occupy our watch on deck would have savoured to
Captain West of effeminacy.

Coming up the coast, too, in the cloudless nights of
these latitudes, he taught us to know the stars. Most
fascinating I found this, much more enjoyable than
fiddling about with exasperating ends of rope. The
North Star and the Great and Little Bears were at this
time below the horizon, but we were already familiar
with them and there was the whole of the southern
heavens to explore.

The finest constellation is undoubtedly the Southern
Cross, the four stars of which would be perfect were it
not that one of them is not so bright as the others.
According to the legend, the Southern Cross was once
visible in the northern hemisphere but in consequence
of 'man's first disobedience' the human race became
unworthy to behold its transcendent beauty, and at the
same moment as the angels took up their stations with
flaming swords at the gates of Paradise, the constellation
was removed to the uninhabited polar regions of the south.

A star-atlas was lent us to study in our watch below
and by its help we made our first excursion into the

absorbing science of astronomy. Little by little we gathered, as I liked to think, 'from the starry fields of night a harvest of divinest thought,' and also, as the old man pointed out, learnt what would be of use to us when we came to sit for our Board of Trade examinations.

The constant south-easterly direction of the Trades made it essential that we did not overrun our destination, for this would have necessitated a weary beat to wind'ard and wasted many days. In consequence the old man stood a bit more in to the coast after passing the islands of the twin saints, Ambrose and Felix. Our first landfall was the slender snow-peak of Misti Volcano. The land at its base was not visible; it stood up, a dazzling triangular white cone, out of the almost-imperceptible haze, a handsbreadth above the horizon. One glance at it was enough to establish our position; there is no mistaking the shark's-tooth crest of Misti for any other peak. It is nearly 20,000 feet high and, I should imagine, was almost two hundred miles away when we first sighted it. We stood in no nearer to the coast, for the breeze was lessening and fined away gradually as we slanted to the nor'ard.

The weather grew very hot as we drifted, rather than sailed, over an almost windless sea. The current helped us a little and next day we were in the latitude of Pisco. The coast was not visible, but at times we could see glimpses of a great wall, lying cloud-like above the horizon. It was the summit of the coastal range of the Cordilleras.

147

Sharks put in an appearance and we caught several. The first one we hauled aboard was a female, for it had eight young ones inside it. These youngsters were perfectly formed and exactly like their parent, but on a smaller scale. When we cut the mother open they slithered out on deck, all very much alive and each about two feet long.

A great difference of opinion arose concerning these young sharks. Some of the crew said they were 'unborn' and had never yet left their mother. Others said they had been duly born, but had swum back down the maternal throat at the sign of approaching danger. I was never able to discover the truth of the matter, but, of the two hypotheses, inclined to the latter. For sharks in embryo they were too much alive and too exact replicas of the big one. There was no indication of 'growing to the head' as children and most animals seem to do, and, besides, I always thought sharks spawned like other fish, and did not bring their offspring into the world like mammals. But on the other hand, to swim down a shark's throat and be cast up again sounded a bit far-fetched and savoured too much of Jonah and the whale. Anyhow, the men cut them up and cooked them for supper, but in view of the doubts as to their nativity I gave them a wide berth.

Next day, to a three or four-knot breeze, we stood in for the land and made the island of San Lorenzo, at the mouth of Callao Bay, just before noon. It was dark before we passed the westernmost end of the island and, drifting very slowly, came to an anchor in the roads an

hour or two later, on our fifty-first day out from Newcastle – not a bad passage considering the days we had
lost in the Cook Straits and our slow drift up the
American coast.

We could see the lights of many vessels, both steam
and sail, anchored all around us in the Bay, and a bright
illumination ashore marked the position of Callao itself.
We turned in that night with pleasurable anticipations,
for here we were in a really foreign country – and a
most interesting one at that – and we looked forward
with eagerness to a closer acquaintance with it on the
morrow.

CHAPTER EIGHT
The West Coast — Callao
*

To some apprentices all the time, and to all apprentices
some of the time, the whole western seaboard of South
America — the West Coast, as it is commonly called — is
a barren region of rock and sand, on the backmost edge
of beyond, uninteresting to a degree and only created as
a sort of penal settlement for unfortunate seafarers.

Certainly it is this aspect which is most commonly
on top to those who come to it standing in under tops'ls
from the Pacific. A coast of open roadsteads it proves to
be, where the sun is always hot and no cooling shower
ever falls, where in many places ships roll unceasingly,
gunn'les under, to the long, unending swell, and where
above all, crews work their own cargoes. One suffers
from heat, drought and hard work, and curses the coast
with whole-hearted fervour and thoroughness.

Suffering such inconveniences, it is easy to overlook
the fact that this West Coast is the Peru of Prescott's
pages, the battleground of Inca and Castilian, where in
later years those picturesque ruffians, the buccaneers,
wrote crimson foot-notes to history. To any one who
cares to search, these latter aspects of the land are plain
enough, but, as I say, under the discomforts of life on
the coast they are apt to be overlooked.

But I am getting a little too far ahead. On that first
morning in Callao we were merely able to take stock of
our immediate surroundings.

THE WEST COAST-CALLAO

When we came on deck at 6 o'clock we found ourselves anchored about three-quarters of a mile from the shore, with the magnificent Bay of Callao stretching all round, its waves a-sparkle in the first beams of the morning sun. A large number of ships, both steam and sail, were moored near, while the white walls and numerous church-towers of the city gleamed pleasantly in the sunshine over the port bow. The town is situated on a low strip of land lying between the mountains and the sea; a few miles inland is Lima and at the back, running north and-south, rises the mighty wall of the Cordilleras, the seaward range of the Andes.

A magnificent sight the mountains were, towering high above the foothills and the plain. Their flinty sides appeared destitute of even a blade of grass; clouds drifted, high up, athwart their face; and above shot up the everlasting peaks. I never grew tired of watching them, weaving day-dreams of adventures in their depths or planning the escalade of some sunset-challenging crest.

The Customs launch soon came alongside and pratique was given us. The agent and comprador followed hard after the Customs, and to our disappointment we learnt we should not be going into dock, but were to discharge our cargo into lighters in the Bay. We heard, too, that bubonic plague was rampant ashore and among the shipping, but this gave us little concern. Plague or no plague, we were eager to investigate the place.

Our first duty was to unbend the sails and thereafter

to rig the gear for discharging the cargo. A lighter came off next morning and we began the long task of unloading.

At this time on the West Coast ships' crews discharged their own cargoes. The engagement of stevedores was not compulsory under the port regulations and, as it was cheaper to employ the crews, the choice was not difficult to make. In the *Arethusa* we had no 'donkey,' everything had to be done by hand and by means of simply-constructed hand-winches – 'Armstrong's patent,' as they were called. Four men were stationed at the winch, four men in the hold to shovel the coal into baskets, the second mate at the gangway to tip the stuff, the mate at the weighing-machine on the bulwarks, and the work began.

We worked from six in the morning till six at night, with half-an-hour off for breakfast and an hour for dinner. The sun was hot, the coal dusty, the baskets heavy, and by the time our day's work was done we had had about enough of it.

To encourage us in our labours the old man laid in a stock of 'pisco,' a cheap, fiery spirit, very popular in Peru. We got a tot apiece every day – two or three when we worked exceptionally well. Sixty to eighty tons was our average day's output, the quantity depending on the supply of lighters.

Our work was not lessened by the fact that the four hands we had shipped in Newcastle ran away on our second morning in port. They cleared out in a particularly bare-faced manner. A shipping-master's boat pulled alongside and was peremptorily ordered off by the mate. The rowers gave way, apparently to obey the order, but in reality to pull under the bows. Our four men coolly lowered their bags and themselves into the boat and were rowed ashore, while the mate, completely unconscious of what was happening, carried on with the work on the other side of the deck-house.

The skipper took their desertion very philosophically. We expected to be a considerable time in port and seamen were plentiful ; meanwhile the remainder of us could do their work. Beckett was made night-watchman and Gilroy and I took charge of the boat. When not engaged therein we were put on shovelling coal, an arrangement we cordially disliked.

We had some compensations though. A new president was elected the week after our arrival. The actual ceremony took place in Lima, but Callao was all beflagged, and the inhabitants made the occasion an excuse for a 'fiesta.' All the statues on the plaza were hung with flowers ; the fire-brigade, with flowers in their caps, paraded the streets, dragging their engine with them; and the wine-shops did a roaring trade. The population turned out to a man and everybody cheered at the slightest provocation, while bands, uniforms and the national colours were everywhere. Gilroy and I were ashore all day and watched the fire-brigade. They marched up and down and round and round with indefatigable enthusiasm. Every time they encountered the national flag the whole array saluted, and as there were two or three of such flags hung out of each building and positive festoons of them at the street corners, they were kept busy.

The warships out in the Bay, and there were six of them, were dressed for the occasion. They each fired a salute of fifteen or twenty guns and their bands kept playing all day. As it was a general holiday the captain couldn't get lighters and no coal was discharged. Having no use for 'fiestas' in the foc'sle of the vessel he commanded, he thought it a good opportunity to overhaul the mooring chains, and our unhappy crew chipped and tarred all day to the strains of stirring music and sonorous metal blowing martial sounds.

Amends were made on the following Sunday, when all hands were given a sovereign apiece and allowed the day's liberty ashore. Gilroy and I had to pull them into the mole in the morning and at eight o'clock in the evening go back for them. It took us two hours to collect the whole hilarious party and the boat was almost swamped coming off, but we managed it without mishap.

The following Sunday the bosun and I succeeded in getting the day off and went up to Lima together. It was Stedman's idea; he spoke a little Spanish, he said; and I hailed the suggestion with delight. We caught an early train and after a pleasantly novel journey of only a few miles reached the capital by ten in the morning.

On arrival we found the city gaily decorated. A 'fiesta' was in progress, consequent upon the election of a new president. The Limenos, good people, were apparently making a three weeks' holiday of it, and the bosun and I, nothing loath, sailed in to help them celebrate.

The streets were full of people – Spaniards, negroes, Indians, and every possible combination of the three. The colour of their faces ranged through all hues from pasty white to jet black, and their attire was as varied as their features. Good-humoured and light-hearted people they seemed and very friendly. What struck me at once was the amazing politeness of everybody. I was inclined to be rather contemptuous of it at first, but altered my opinion on longer acquaintance and came to admire the Spanish character.

We strolled round the Plaza Mayor for a while and had dinner at a restaurant, its front shaded with palms and open to the street – very cool and pleasant. Our appetites seemed to astonish the waiter. We finished the dinner he brought us and paid for it, then ordered another. The good man stared hard, but duly brought it. There is a persistent belief in South America that all Englishmen are mad and, in sympathy with our affliction, the waiter charged us only half as much as for the first helping. We finished that off too. 'I could go a bit more,' said the bosun, 'we don't often get a feed like this.' I agreed with him and we hailed the waiter. 'Dos mas,' said Stedman, holding up two fingers, 'mucho hungry, hombre!' The waiter seemed perturbed; he ripped out with a heavy broadside of Spanish and what sounded like a volley of imprecations, but went away and returned with two huge platefuls, which he set down before us. I hauled out a handful of silver but the waiter shook his head and fired off some more Spanish. We pressed him, but he persisted in his refusal, so we finished our dinner and walked out into the street, leaving a handful of centavos on the table, to convince him we weren't a couple of destitute mariners who hadn't had a meal for weeks.

We went to some races in the afternoon. It was the first public appearance of the President and the race-course was a very gay sight. The number of soldiers and policemen about was astonishing. Very brilliant fellows they all were, and armed to the teeth, if a little dilapidated in the matter of boots.

Coming back we passed through the Chinese quarter. There are Chinese in every town on the Pacific, for your almond-eyed Celestial is above all things a merchant, and if his social qualifications are not extensive, he creates wealth wherever he goes, and is tolerated if not welcomed. I forget if I've mentioned that Stedman had for some years been skipper of a tug-boat on the West River, and spoke Chinese — the Canton dialect — fluently. It is a difficult language to learn and very few white men master it. Anyhow, on our way back, we went into a little grocery-store, and he addressed a wizened old Chinaman sitting behind the counter, in his mother-tongue.

I never saw anybody so startled. The old chap jumped up, jabbering nineteen to the dozen, and peered hard at us, then screamed into the back of the shop, and immediately a whole flock of little Chinamen ran out, gesticulating and jabbering. They surrounded Stedman, all speaking at once and seemingly wildly delighted. The bosun stood there, smiling, answering as fast as he could the heathenish clatter of each tongue in turn. Those Chinks almost embraced him, they grinned in a way I have never seen Chinamen do before or since, and when we left they pressed upon him several little packets of tea and weird Chinese delicacies. They followed him to the door all talking and laughing, completely shaken out of their native Sphinx-like immobility.

'What on earth was it all about, bosun?' said I, when we got outside again.

'Oh, I saw they were Canton boys,' he said, 'and I talked to 'em about the place. They hadn't been there for years, and wanted to know all about it, but they couldn't make out who I was at all.'

Altogether, our day was a great success, and when we got back to the ship at eleven that night I determined to see more of Lima, if possible, during our stay in port.

I only had one more opportunity, however, and that was with the captain just before we sailed. It was a different sort of visit to the previous one, but no less enjoyable. We went over the great Cathedral that occupies one whole side of the Plaza d'Armas, the twin towers of which were plainly visible from the ship. It is a magnificent building, and, after the famous Cathedral of Mexico, said to be the finest place of worship in the New World.

The Cathedral is but the chief of the many fine churches in Lima. Most of the public buildings, too, are

imposing structures, the streets are wide, with avenues
of trees, and altogether the place well deserves its name
of 'the city of magnificent distances.' The architecture
of nearly everything is immensely solid; the domestic
type particularly so; each house looks as though it were
built to withstand a siege. Most of the better-class
residences are in the form of a hollow square, with bare
thick walls, facing the street, looking inward to an
enchanting courtyard filled with shrubs, flowers and
perhaps a fountain, and surrounded by balconies. The
Spaniard got the idea from the jealous Moors, and it
suits well with his own reserved character, besides
making a very snug homestead.

But these occasional jaunts were only interludes in
the steady task of discharging our coal. No letters from
home reached us; the books we had purchased in the
Colonies had been read and re-read. A Bible and my
water-logged Shakespeare were our only literature,
and when those palled we turned in.

Big Mac had purchased some needles and Indian ink
in Newcastle, and spent his time tattooing. We all
patronized him, and some of the men were worked all
over like Maori chiefs. Mac had a great variety of
designs, which he put on by means of transfers and then
pricked in with five tiny needles tied in a row, using
green, blue and scarlet ink. Most of the men were
already tattooed – ships, snakes, eagles and clasped hands
being the favourites – but old Jamieson had a tomb-
stone bearing the motto 'In memory of Mother' on
his forearm – a strange device for an old shellback, as I

then thought, though I have since seen similar ones quite a number of times. Neilsen, the Norwegian, had a design from his hip to his ankle representing a sword worked, as it were, in and out of his flesh, and big Mac himself was a medley of anchors, frigates, bleeding hearts, and barrels of rum.

We were able to do a little ship-visiting now and again, for there were a number of sailing-ships in port. The *British Isles*, a big full-rigger, lay next to us. Her second mate came aboard one evening to inquire if we had seen anything of a man who had jumped overboard the previous night in order to desert. We had not, and in view of the number of sharks that infest Callao Bay we thought it unlikely anybody had, and told him so. The *Eva Montgomery* was another vessel that lay near us. She had most beautiful lines and was reputed a very fast ship. We met her again in the Channel on the homeward passage; she was carrying royals to a stiff breeze that whitened the short Channel waves and made as pretty a sight as one could wish to see.

One evening we had a little excitement on board. Tommy had brought the shark-hook along during the day to try for a big brute that had been persistently hovering around for some time. We were having tea in the half-deck when we heard a terrific yell from Tommy and, rushing out, saw him holding desperately on to the line. It was being dragged helplessly through his hands. We rushed to his assistance, so did Mac and some of the men who were standing for'ard. Gilroy dashed aft to call the old man, and the latter came up wiping his mouth with a napkin and grabbing a harpoon from under the port rail as he came along.

Half a dozen of us were hanging on to the line and having the deuce of a job to hold it. The fish was a monstrous big one and cunning too, for he kept making rushes from side to side that slackened the line for a moment and then almost tore it from our grasp. We could see nothing of him, for there were about ten fathoms of line out. 'Steady boys,' said the old man as he came up, 'keep a steady strain on the line. That's not a shark.'

The skipper climbed on the rail with harpoon ready to strike. We got the line in inch by inch, the fish, whatever it was, being as strong as the half-dozen men fighting him. Suddenly we saw a great commotion in the water twenty feet away; what looked like two great grey wings beat the surface and a broad, black back appeared.

'Sting ray,' said the old man, 'pull, boys!'

We hauled a few more feet of line in, then came the captain's opportunity and we saw him drive down with the harpoon. He hit the fish fair and square and, jumping off the rail, hauled away on the line, more of the men tailing on behind him. We hauled the beggar almost up to the ship's side, then, making the harpoon-line fast to the cargo-fall, took it to the winch and hove the ray up bodily.

It swung in and came down on deck with a mighty flop, all hands jumping clear. It was a hideous monster, mottled black and grey, with its body prolonged into

flat, wide wings on either side, and with a whiplash of a
tail, six feet long, no thicker than one's finger and armed
with needle-pointed spikes. The harpoon had cut a
wicked gash right in the middle of the creature's back;
the wound bled like ten bullocks and the blood ran out
of the fore and main scupper holes and made that side
of the deck look as if we had been in action. We
measured it and found it was nine feet three inches from
wing-tip to wing-tip and seven feet six inches from the
head to the base of the tail. It was about eighteen inches
thick and must have weighed nearly 400 lbs. The flesh
was good for nothing, so, after hacking out the harpoon
and hook and cutting off the curious tail as a memento,
we dumped it.

This was the largest sting-ray, or 'devil-fish' as they
are often called, I ever saw caught. They are portent-
ously ugly brutes, with a long gash for a mouth, situated,
not at the end of the head, but underneath, something
like that of a skate – which the sting-ray closely
resembles. They are prodigiously strong, as we found,
and I have heard it said that they are able to jump clear
of the water by the aid of their wings and flop along in
and out, like a stone making 'ducks and drakes,' though
I never saw one do it. They are cannibals in their
habits, for a subsequent specimen we caught and
cut open proved to have parts of another one in its
stomach. Their long, spiked tails too, are dangerous
weapons; and altogether these big sting-rays present a
remarkable congruity between their appearance and
their appellation – sea-devils.

A GİPSY OF THE HORN

Whether the skipper's sporting instincts had been aroused by the capture of the ray or not, I don't know; but next day he had the gig fitted out with a yawl rig. The masts and sails had been specially made for her under his direction, on the passage out. The gig was a Bombay-built boat, made of teak, and looked a very dainty little craft in her sailing rig.

On one occasion we sailed over to San Lorenzo, the island that shelters Callao Bay to seaward, and made a landing on its shores. We picked up a few shells there and saw a number of large crabs. The old man said they were deep-sea crabs and often met with hundreds of miles from land. They were big, vicious creatures, a dull, mottled green in colour, with most business-like pincers. They scuttled about rapidly and differed from our English crabs in that their shells and claws were flattened, with sharp, serrated edges. Their shape enables them to swim very quickly and it is surprising with what speed they can move along just under the surface of the water, working their legs as though they were walking on terra firma.

A few days afterwards we had a more ambitious excursion. We sailed down to Pachacamac islands, starting early in the morning and taking a supply of food and water in the boat. Pachacamac lies to the south'ard, a full fifteen miles from the anchorage, and is the name given to two little rocky islets and a ridge of reefs not far from the coast. It is a wonderful place for birds and fish. There were numbers of octopus there too — there are round most of the barren islands and rock-studded bays of the West Coast.

By the 3rd of October we had discharged about twelve hundred tons of our coal. The space under the main hatch was clear, so we stopped unloading for a time and began to take in a couple of hundred tons of stone ballast. The question of where we should go to load cropped up and we discussed it at length. Some thought 'Frisco, others the East Indies or perhaps back to Newcastle, but the general opinion seemed to be that it would be somewhere down the coast to load guano.

Guano and nitrate are the two chief exports of Peru and Chile. The latter is loaded at ports further south, Iquique being the most important. Guano comes from a number of promontories and islands scattered along the coast between Pisagua and Los Lobos. The Chincha Islands off Pisco are about the best-known. When first discovered these small barren rocks — for they are no more — were covered with a thick layer of the stuff, over a hundred feet deep. They are almost worked out now and not many vessels go there to load, the Peruvian government having placed restrictions on the indiscriminate working of the deposits.

The absence of rain on the coast prevents the chemical properties of the guano from being washed away and has created an enormous fund of wealth. Millions of tons have been removed from the various islands, and fresh deposits are constantly being laid down. Birds of all sorts, but particularly pelicans, haunt the coast in such myriads that frequently the air is darkened

163

by a vast impenetrable cloud of them flying overhead. To give some idea of their almost unbelievable numbers, it is estimated that half a ton of guano is deposited every year on each square yard of a guano island. And this sort of thing prevails for some hundreds of miles along the Peruvian seaboard.

But guano or any other commodity – I personally had small choice in the matter and cared little where we went to load. All places were new to me and I was only anxious to be quit of the coal and see the barque trim, clean and at sea again, with a ripple overside and her royals set.

One day as we were working at the ballast we experienced a slight earthquake. It was not a very severe shock; on the ship it felt as though we had collided with a lighter, and ashore it caused no consternation. On our next visit a much slighter shock was sufficient to put the inhabitants of the coastal towns in a panic.

But these irregularities did not stop the even tenor of our discharging. On the 18th of October we finished our coal and mighty thankful we were to see the last basket swing over the rail. As Gilroy and I pulled the captain off in the boat that evening I asked him where we should be going next.

'We're bound for Guam,' said he, as he climbed up the gangway.

'Guam? Guam? – where the blazes is that?' we asked one another. As soon as we got along to the half-deck we turned up our old atlas and found the place – a little speck of an island far away in the Ladrones.

What on earth are we going to do there, we wondered.

After tea I went along to tell the bosun of our destination. When I mentioned Guam he laughed:

'Yes,' he said, 'the crowd for'ard always is bound for Guam. Did the old man tell you that?'

'Yes,' said I, and then Stedman explained.

'Bound for Guam,' turned out to be an old sea-term meaning that the vessel's destination is to be kept secret. It completely took me in and the others too; the bosun and Jamieson were the only men for'ard who knew what it meant. When I inquired of the former as to the origin of the term he couldn't tell me and I have never since been able to learn.

For the next few days we were kept busy getting in the remainder of our ballast, bending sail and making the ship ready for sea. Our water-tanks were filled and four new hands came aboard in the place of those who had run on arrival.

One of the latter was a very queer old customer. His name was Fox and he had been chief mate of the *Cragenan* but was discharged for incorrigible drunkenness. He was a small, wiry old veteran, with a deep, gruff voice and an exasperating trick of repeating verbatim every order given him, with a hearty 'Sir!' tacked on the end. Of course it is only seamanlike to repeat an order, but not with the thoroughness that Fox employed.

'Aft here, Fox,' the old man would say, 'and haul taut this sheet.'

'Aft here, Fox, and haul taut this sheet, SIR!'

'Lively now, man, get a move on.'

'Lively now, man, get a move on, SIR!'

'Damnation! the sheet, you old fool! Leave that tack alone!'

'Damnation, the sheet, you old fool, leave that tack alone, SIR!'

It drove the old man almost frantic; he could hardly tell Fox never to repeat an order, so he cursed all the more and back, with clockwork precision, all his curses came, a gruffly-polite 'Sir!' dangling behind them. We quite enjoyed hearing the skipper swear at Fox, but we had to keep our enjoyment to ourselves.

When the new hands came aboard word went round that we were bound to Santa Rosa, a small guano-island in Independencia Bay, some distance to the south'ard of Callao. The rumour proved correct, and at mid-day on the 28th we hove up anchor and got under way. We had been nearly eight weeks in port and during that time had shovelled and hove up by hand about two and a half thousand tons of coal and stone. Pleasant as parts of our stay had been it was entirely without regret I saw the headland of San Lorenzo slip by, as we stood out through the Boqueron under our own sail and, soon afterwards, picked up the first breath of the south-east trades.

Independencia Bay, our destination, was less than two hundred miles, as the crow flies, south of Callao. Owing to the direction of the prevailing winds it was impossible to shape a direct course thereto. The South-east trades blow incessantly along this part of the coast and make it a difficult matter for a sailing-ship to win only two or three miles of weatherboard. The Peruvian or Humboldt current also runs up the coast from south to north, following the trend of the shore, and adds an appreciable amount of leeway to a vessel's track. These two factors combined make it impossible for a sailing-ship to steer straight for her objective if it lies due south.

Such being the case, we had to stand away on a long slant, close-hauled to the trade-wind on the port tack, keeping course until we were down almost to the latitude of Juan Fernandez. In about 30° South we went about, braced sharp up on the opposite tack and ran down to windward of our destination; then, going large with the coast in sight, stood in for the anchorage.

We were thirty days on the passage – a solid month of hard sailing in fresh, breezy weather. During the time we covered a round two and a half thousand miles making good that hundred miles of weathergage.

Such are the limitations of sailing-ships, the steamboat enthusiast will say. The remark is just and I grant him his slippery heels and extra turn of speed. Too well I know that the stresses of the present day must inevitably eliminate such old-fashioned methods of travel as clipper ships. Oars have gone, sail must go, and God only knows what will take the place of steam.

Steamers have it all their own way just now and one of them, I admit, would easily make the passage from Callao to Santa Rosa inside twenty-four hours. Yet is there something to be said for the out-of-date windjammer. If the simple joy of living is what makes life worth while, and not merely the amount of work and worry crowded into the twenty-four hours, then assuredly the old canvas wings cannot pass without a sigh of regret. Was the life hard? — one slept the sounder for it. The food poor ? — salt pork tastes better to a man in Fifty South than a banquet to an epicure. The pay paltry? — there is other wealth, thank God, than that of gold.

The sailing-ship was an exacting mistress to serve. She was all that; she was a heartbreaking wench at times, yet none the less a Cleopatra among the sisterhood of the sea, inspiring an affection the ladylike liner is powerless to evoke. Her passages were made by the sea-cunning of men, not by the strength of machinery. Her sail-plan had the complexity of a woman's character and, woman-like, 'her infinite variety' would respond to every thought and act of one who understood and loved her.

Twenty-four hours and we took a month! Let it go at that. I stand to it — there are few better ways of spending a month than on board a heeling clipper, running down the south-east Pacific trades.

Throughout the passage we had splendid weather. With every stitch set we bowled pleasantly along, leaning over at a steep angle and making a fair amount of leeway, for we were flying light and showed a side out of the water like the wall of a church. We encountered a fair sprinkling of the squalls for which the Pacific is notorious — little puffs that sent us rushing to the halyards and constrained us to let the royals and upper stays'ls fly.

One morning as I was working on the foc'sle head I saw a curious sight. Far ahead on the port bow I could see a prodigious commotion going on in the water. I had several glimpses of a great black arch suddenly rising out of the sea. The object rapidly drew nearer and as it did so I saw it was a whale. Every minute or two it would leap out of the water bodily and fall back, not head-first in a gliding curve, but flop! just anyhow. It hit the water each time with an almighty splash and a noise like the report of a gun.

The bosun came up on the foc'sle head and looked at it. He said he thought the whale was being attacked by 'killers' or blackfish. 'Blackfish' is the sailor's name for the grampus, a kind of small whale itself. They are met with cruising about in schools and are creatures to be avoided, for they yield next to no oil and are very ferocious.

On this occasion we didn't see any near the whale, though they swim with a black dorsal fin curving in and out of the water which makes them very conspicuous. In this case it may have been because the whale himself was kicking up such a commotion.

He was going the opposite way to the ship and soon swept past and was lost to sight astern, leaping, writhing and splashing to the last. I think the bosun must have been right in his conjecture, for there was evidently something very wrong with that whale.

This was the only incident on the passage worth

recording. On the 24th November we came in sight of the mainland again, a little to the south of Ylo. We stood rather far in to make certain of our position and so lost the benefit of the trade-wind, which is usually light for a distance of a few miles off shore. As a rule we kept a good offing, but in this case the skipper was anxious not to overrun his port and be under the necessity of making another long beat to seaward. The very proper dislike of sailing-ship men to hug a coast too closely, with the possible danger of finding themselves on a lee shore, is responsible for the modern jest that as soon as a windbag smells the land it's 'Ready about! Let's get out of this!' Cutting corners is all very well in a vessel that has a perennial fair wind under her counter in the shape of a propeller, but no shipmaster who has had to claw off a dangerous shore under sail wants to repeat the experience gratuitously.

A bare, desolate and mountainous coast this of Peru is. The coastal ranges are treeless, waterless wildernesses of rock and sand. They stretch away in blotches of yellow, purple and brick-red to the far-distant grey wall of the Cordilleras. No building, no trail of smoke, no vestige of human life relieves their stern desolation. The ragged coast-line is endlessly a-boom with the thunder of surf and fronts the sweep of an unbroken seaward horizon, while over all lies the shadow of the stark Sierras. A romantic land, I thought, for the great empire of the Incas, and no unfitting theatre for the exploits of Pizarro and his Spaniards.

For two days we coasted slowly along to the nor'ard

and on the third day, early in the morning, as the rising sun sent a line of gold rippling along the distant edge of the Andes, sighted the mountainous promontory of Mt. Quemada right ahead.

This headland, a towering block of stone over 2,000 feet high, falls sheer into the waters of the Pacific and forms the southern end of Independencia Bay. The island of Santa Rosa, our destination, a small flat platform of rock only a few hundred acres in extent, is on the opposite side of the entrance. The anchorage lies behind the sweep of Mt. Quemada, between the mainland and the island, and is approached by a very narrow entrance – the Serrate Channel.

Disaster almost overtook us as we headed slowly into the fairway. The wind had fallen very light and was only just sufficient to ruffle the surface of the long undulating rollers. The royal yards had been lowered but all the rest of our canvas was set to waft us along, when, without a moment's warning, a violent gust of wind swept down the slopes of Quemada, filled our sails and urged us through the water, straight towards the seaward-stretching reef that fringes the island of Santa Rosa.

For a few moments there was the deuce to pay. T'gallant and tops'l halyards were let fly and the yards came down with a run, with the canvas flapping in the breeze.

'Let go!' yelled the old man to the mate on the foc'sle head. Chips sprang to the windlass, while the bosun seized a topping maul, jumped for the cathead and

knocked out the pin of the ring-stopper. The anchor
went down with a rush and a roar and the cable leapt
out so furiously that the windlass-brakes caught fire.
Old Chips was almost blinded by the flying sparks and
dust, but hung on to the brakes and gradually checked
the mad rush of the cable. The anchor held, and as we
clewed up the sails in desperate haste, the barque's way
slackened and she came to a standstill within a biscuit's
throw of the reef.

The puff of wind only lasted a couple of minutes. It
was all over in that time and was succeeded by a dead
calm which enabled us to warp the ship into the fairway
and moor her properly in the anchorage, with both
bowers down and thirty-five fathoms out on each.

These sudden flaws of wind, we soon came to realize,
were one of the peculiarities of Independencia Bay.
They swept down the mountain-side without warning
and caused us no little inconvenience. In particular,
they made boat-sailing a highly dangerous pastime. The
Peruvian name for the bay, among the lancheros that
is, was 'Bahia de viento' – Windy Bay.

CHAPTER NINE
The West Coast – Santa Rosa
*

WHEN the ship had been snugly moored and the sails furled, we had time to look about and admire the spot to which Fate had sent us. It was an open roadstead, and the entrance lay clear and wide to the long Pacific swell, causing us to ride with a constant grind and heave on our straining cables. In the event of a gale from the south-east we should have to slip and run, so sail was left bent on the yards in readiness.

The bay itself is only a mile or two in length and less in width, and is formed by three islands – Santa Rosa, Lobos and the Morro de Viejas – lying between Mt. Quemada and Carretas Head on the mainland. It is a wind-swept, rock-littered trap of a roadstead, one of the wildest and most outlandish places, I thought, that a Christian keel could well be brought to anchor in. In bad weather – and for a full half of the time we were there it was bad weather – the place is a regular nightmare and calculated to sprinkle grey hairs on a shipmaster's forehead. The Admiralty charts give some idea of its shape, but none at all of its inexpressibly savage appearance.

Ahead of us, seen through the narrow entrance of the bay, lay the tumbling expanse of the Pacific. On our port bow rose Mt. Quemada, a stupendous mass of rock – split in ravines and knife-like ridges – treeless and naked to the glare of the sun. South and west the coast

173

trended away from it, a barren, verdureless waste, ever rising in rolling waves and hillocks of sand, with black rocks jutting out here and there, till it merged into the huge, impassable mountains – La Cordillera de Carrasco – that ringed about an arc of our horizon.

Astern of us were the rippling, rock-staked waters of the Bay, while to starboard lay Santa Rosa and its two smaller neighbours – low, flat, rocky islets covered with a rich deposit of guano, with the sandy slopes of Viejas beyond. The islands were absolutely barren, not a stream of water nor a blade of grass clothed their relentless nakedness.

And they call this place Santa Rosa, thought I, as I looked about me – 'the sainted Rose' – spare me, then, the thorns!

The place would have been stern and forbidding for a penal settlement. And the population were in keeping with their surroundings. The guano-workers were a wild, rough lot of half-bred negroes and Indians, sack-clothed and uncivilized. They lived in little tents and lean-to's of sacks and tins constructed on the highest part of the main island. There were about eighty of them altogether, including a Spanish foreman and an English chemist, who were in charge.

All their food and every drop of water was brought to them from Pisco, in a little sloop that made the journey once a week – six hours there and six days back. On her they depended for every necessity of life; if she had failed them, their plight would soon have been desperate.

Built out from the top of the island opposite the ship

174

was a rough, sloping pier down which the sacks of guano
were slid into the lighters that anchored at its extremity.
There were four of the latter, all very dilapidated old
craft, and when not in use they found such shelter as
they could to leeward of the rocks on which the pier
was built.

The sack huts, the pier and the bumping lighters
were the only signs of man's handiwork between sea and
sky. In their rough uncouthness they only deepened
the impression of wildness and desolation. The combination of mighty mountains, barren, sun-scorched
rocks and rolling sea was wild and grand in the
extreme, and fascinating by reason of its very untamed
savagery.

Yet desolate as it was, it was the *splendour* of desolation; there was no suggestion of anything in a minor
key about it. The colour scheme was bold and blazing –
a vastness of sapphire and indigo the sea and sky, of
black and blinding yellow the towering, tumbled land.
Sunrise every morning was a thing to stand in silent
wonder at, as stainless peak and far-flung desert flushed
a rosy pink and seemed to quiver in the delicate, dewy
air. Sunset was no less impressive – a blood-red orgy of
triumph and a flaunting of golden banners in the western
halls of the Pacific.

The morning after we arrived the Spaniard and his
colleague came off in their boat and had an interview
with the captain. One result of their conference was
that we were ordered to get the stream-anchor out and
moor the ship by the stern.

The old man borrowed a lighter for the operation, and a gruelling day's work it gave us. We first took the lighter for'ard and lowered down into it the stream-anchor, with forty fathoms of mooring chain shackled on. The anchor was slung over the lighter's stern, which was then dropped aft some distance from the ship, and the anchor let go. The end of the chain was passed aboard and hove taut by means of wire runners taken to the capstan. It was hard, dirty work, danger-ous, too, in the heavy swell, and took us till dark before the ship was moored stem and stern.

Our first lighter of guano came off early the following morning, and, rigging the cargo gear, we started on our two months' task of loading the ship.

The crew, under the direction of the officers, had to do everything. The lancheros were a lazy lot of beggars, and would do nothing after they had made their lighters fast alongside. We had to sling the sacks of guano in the lighter, heave them on deck, empty them down the hatches, and trim the stuff in the holds. It was not possible to work every day; bad weather at once put a stop to the operations, but when conditions were favourable all hands worked from daylight to dark.

I soon realized guano was not a pleasant cargo to handle. In fact, it was heavy, filthy, clayey stuff, full of the decomposing remains of birds, and smelling so powerfully of ammonia that, later on, when the holds were getting full, it was impossible to work down below for more than a few minutes at a time. We tied vinegar bandages round our mouths and nostrils, but, even so, the pungent odour penetrated to throat and lungs and drove us gasping on deck after a few minutes' spell at shovelling below.

Yet in spite of the horrible stuff that was daily more bemiring our trim little barque, the time passed by no means monotonously or unpleasantly, at least for us in the half-deck. About every third day was a 'surf-day,' and wind and sea for a time put a stop to all loading. On these occasions the old man ordered all four of us into the gig, and we set out for day-long shooting and fishing expeditions up and down the coast. We had some fine outings. We used to put a supply of food and water in the boat, the skipper brought rifle, harpoon, fishing-lines, nets and dynamite cartridges with him, and off we went. We were frequently at the oars from six in the morning till eight at night, and had some splendid trips.

Then there was the arrival of the sloop, which always provided a little mild excitement. She went to Pisco and back, taking a week on the trip, in consequence of having to beat homeward against the south-east Trades. On her regular appearance depended the existence of the little community ashore. A sharp look out was always kept for her down the Bay, and, as soon as her anchor was let go, work was suspended and everybody bore a hand transferring the water and provisions from her hold up to the camp.

The little sloop was not always up to time, and more than once the Spaniard had to come off and borrow water from us. The grudging barrel he used to get was taken ashore with a care and solicitude that the live bullock the sloop brought never received.

We ourselves had to be very careful with our water supply. All hands were put on a strict 'whack' and not a drop was allowed for washing purposes. Salt water and soap, we found, do not work well together, and our personal appearance suffered in consequence.

At the end of the first day's work, Jimmy determined to have a bathe to clean and refresh himself. He dived off the gangway and swam round for a few minutes, but as he clambered aboard he ran into the second mate. That officer stared at him for a moment, then opened his mouth and swore steadily and fluently for ten minutes. Between the imprecations Jimmy gathered that the Bay was swarming with sharks, that he was a something – something fool, and that it was the second mate's unalterable conviction that the devil looks after his own. Jimmy got away at last, but there was no more bathing while we were at Santa Rosa.

We had a very heavy blow about a week after our arrival. The sprays from the huge seas that broke on the seaward reefs made a clean breach of the island, and little cascades of water trickled down from the cliffs into the Bay. The furious wind and sea did a good deal of damage. Most of the rickety wigwams on the island were blown down, a couple of lighters drifted ashore, and were smashed up on the rocks, and our gig, which was lying astern, was almost swamped at her moorings.

The Governor of the island had a pretty little yawl, about the size of a ship's life-boat, which he used for pleasure cruises. He and eight Indians were out at sea in her when it came on to blow. They beat in to the island, but the sea was running too high for them to land, and they were blown out again. For two days they fought against wind and sea in the offing, with neither food nor water, till the wind moderated and they struggled in again. It was a haggard, storm-beaten party that finally scrambled ashore; we had pretty well given them up for lost.

We put the lifeboat over the side to succour a lighter that was in distress, and eventually towed her to the ship and made her fast astern. That was the only time we left the ship; boating-trips were out of the question while the blow lasted..

The following Sunday — every Sunday, in fact, and very often on week-days as well — we set off on a day's excursion — the captain and we four apprentices. The first Sunday we pulled out through the entrance of the Bay into the main ocean, rounded Mt. Quemada, and coasted down to the south'ard for some miles, until we came to a narrow gulf running up into the land.

This gulf was one of the most curious places I ever entered in my life. In spite of a fairly smooth sea, the unending Pacific rollers were breaking with considerable violence on the rocks. The captain espied what looked like a passage, and, giving way at his word, we shot through a cauldron of bursting foam, turned sharply

179

to port and found ourselves on the smooth waters of a lagoon.

Roughly oval in shape, with a diameter of not more than a cable's length, on all sides the naked rock ran up in sheer precipices between one and two thousand feet high. The depth of this stupendous pit – for it was no more than a crack in the mountains into which the sea had found its way – plunged its bottom, on the waters of which we floated, into semi-darkness. Our intrusion startled myriads of sea-birds from the ledges and fissures of the precipitous walls, which circled round screaming, in such numbers as to completely shut out what little light there was. The waters of the gulf were studded with rocks, and on these we landed and secured numbers of different kinds of eggs and young birds. There were star-fish and sea-eggs in abundance, and we picked up some wonderfully shaped and coloured specimens of the latter.

We tried to make a landing on the edge of the gulf, but though we made a complete circuit of the place, everywhere the walls ran down sheer to the water, with slime and seaweed at their base, and frustrated all our attempts.

On our way back to the ship we landed on several kelp-covered little islets, through the fine boatmanship of the captain, and at no small risk to ourselves, and obtained a lot more eggs and strange examples of shore-line life. When we returned to the ship that evening we had a small museum in the bottom of the boat – scores of fish of various kinds and a whole pile of eggs, two shags and a cormorant that the skipper had shot, star-fish, sea-eggs, and a whole medley of beautifully-coloured shells and marine weeds.

On another occasion we pulled all round the Bay itself, landing on the two islands adjacent to Santa Rosa and at several places on the mainland. The scenery was magnificent, the glorious breeze and flashing sunshine only intensifying its naked grandeur.

A perfect preserve for fish and bird life were the waters and shores of Independencia Bay. We had some grand sport hunting and fishing, but it was equally enjoyable just to watch the multitude of strange creatures around us.

We caught enough fish to keep all hands supplied, and, in addition, to lay in a stock for use on the way home. These fish we split and dried on the foc'sle head in the sun. They were then sprinkled with salt and put away in barrels. When we left the Bay we had three barrels chock-full, mostly of small rock-fish, which proved a welcome change to eternal pickled beef and pork. Some of the fish we seined for with a net, but the majority we captured by means of dynamite cartridges. We would pull in among the rocks, and the old man, standing in the stern of the boat, threw a cartridge into the water. It exploded just under the surface, and the concussion stunned all the fish in the neighbourhood and they came floating to the surface in large numbers, white bellies uppermost. We clutched and scooped them into the boat as quickly as possible, for the effects of the concussion were not very lasting, and they soon came to their senses again.

Such a wholesale way of obtaining fish has scandalized those anglers at home who have heard of it, and, I believe, is strictly prohibited in most places where men fish. But it is very efficacious, nevertheless, and, as for its legality — there's never a law of God or man holds good on the Peruvian Coast.

Besides fish, the Bay was a wonderful place for birds. There were uncounted millions of them, ranging from the mighty condor down to little shags. The old man frequently took a shot at one of the first-named monstrous vultures sitting contemptuously on some crag well within range. He never killed one, though. The closeness of the condor's coarse black plumage, and the leathery skin that covers head and neck, made them invulnerable to anything less than a big-game rifle. Certainly they proved impervious to his little pellets. Even when fired at point-blank, they often would not take the trouble to move, but fixed us with a cold sarcastic stare.

Other birds frequenting the Bay in large numbers, of which we sometimes shot or caught a specimen, were flamingoes — tall, supercilious-looking fowls, quaintly described by an old navigator as 'so long-Legg'd that they walk through Lakes without wetting their Feathers'; buzzards, the Spanish 'gallinazas,' large as turkeys and ravenous as sharks, that will stay by a carcase and gorge themselves so full that they cannot rise and may be killed with sticks; and penguins —

'child-birds,' as old seamen used to call them from their close resemblance to small human beings, though to my mind, they looked more like little old men than children.

We caught several whole families of penguins alive — father, mother, and three or four fluffy, squeaking children. The little ones we easily captured, and brought aboard to keep as pets, but the parents gave us more trouble. They would retreat, waddling backwards, to a hole in the rocks, squawking and pecking furiously. Once Jimmy unguardedly put his hand in to drag one out, and got an awful peck for his pains. We used to smother them in a sack and bring them aboard, but they wouldn't eat, and we let the majority of them go, though one or two were killed for the sake of their skins. The markings of their plumage are very pretty, white on the breast and a sort of grey mosaic on the back. Their feathers are thick and very soft, and satiny to the touch.

Most common of all the birds were pelicans, the most solemnly-ludicrous fowl a humorous Creator ever made. They are as large as swans, dirty yellow in colour, with coarse feathers and an immense top-heavy beak. When flying, they keep their knob-like heads thrown back, as though conscious of their unwieldy beaks and a little disgusted with them. But they can fish like heroes, and are even greedier than most other seafowl.

It is hard to speak of their numbers without seeming guilty of exaggeration. They haunt the bays and islands of the Peruvian coast in such myriads as to make the report sound almost incredible. One day we saw an immense flight of them, forming a solid phalanx that we estimated at seven miles in length and of impenetrable thickness. They passed over the island so densely packed that the air was darkened beneath them, and the rushing of their wings made a noise so loud we could hardly hear one another's voices.

Besides all these and the types of purely marine fowl — cormorants, divers, gulls, shags, and the like, whose name was legion, there were large numbers of seals and sea-lions, and frequent shoals of porpoises.

The former lay out in the sun, on the little islets with which the Bay is dotted, so thickly that often not a vestige of rock was visible.

There was nothing inappropriate in the name of Santa Rosa's nearest neighbour — Lobos Island.

Seals are inquisitive creatures and they raised their heads to stare — round-eyed — at the boat. If we shouted at them, the whole party would glide and tumble off into the water, going in like great slugs. A moment afterwards their heads would bob up in all directions, and their round, pathetic eyes be fixed on the invaders of their peace. It was very comical to see a score of sea-lions — some of them monstrous fellows twelve feet in length — with bodies half out of the water, staring with wistful intentness at the boat. If we lay on our oars and disturbed them no more they soon returned to their basking-places in the sun, flopping and scrambling up the rocks by means of their fin-like flappers with remarkable agility.

Once we managed to harpoon a seal, and towed it to the ship, where it was hoisted aboard and cut up. The skin was not heavily-furred like that of Arctic seals, but covered with a fine down and very thick and oily. We cut it into strips and used it for chafing gear aloft.

On Friday, the 4th December, we finished discharging our stone ballast – we simply dumped it overside – and had taken in about 500 tons of guano in its place. From then to Christmas we worked steadily away. About five days a week was our average, and we laboured from daylight till dark.

'Knock off' time came at six o'clock, but when the weather permitted the old man got in the habit of ordering off a lighter about five-thirty. Of course, it had to be finished, and necessitated working a couple of hours overtime. The men stood it for a bit, then saw through the dodge, and one evening at six o'clock refused to work any longer that day.

They went aft, and to our surprise the old man took it very quietly, and said but little. Next day sharp at six came the order ' That'll do, you men,' and all hands trooped off for'ard, mightily pleased with themselves and vowing that 'no blasted skipper was going to humbug them about.'

They hadn't got well into the foc'sle before there came the sharp order: 'Man the windlass!'

'Hullo!' we said, wonderingly, 'what's up?' We soon knew. The order was given to slack away aft and then to heave in on our starboard cable. Mighty hard work it was riding to that heavy swell, but in an hour or so we

185

had hove short. Then the cable was paid out again, and we had to heave tight our heavy stern moorings once more. Till past eight we worked, all dog-tired, and longing for our tea.

Being in a dangerous 'open roadstead,' the hands couldn't complain at work 'necessary for the safety of the ship.' Getting a better hold with the anchors, the old man called it, and it would have been mutiny to contradict him.

For two or three nights this went on, and then the lighters commenced to come off again at half-past five. The men worked them out to a finish with never a word said, and apparently at the same moment the captain was satisfied of the security of his holding ground, for there was no more work at the anchors.

It was check-mate, without bullying or hard words. We boys suffered equally with the men, but couldn't forbear chuckling at the skipper's craftiness, even in the midst of our loud complainings. He was a Tartar to catch, was our worthy Commander.

On one of our frequent expeditions we landed on the lofty islet at the further extremity of the Bay and scrambled right up to the summit. A magnificent view it afforded us. We were full in face of the Morro de Viejas, with a sapphire sea beneath, rolling wastes of sand on either side, and a glimpse of far-distant peaks on the skyline.

On another occasion we pulled out through a narrow channel between Santa Rosa and Corcovado to the seaward side of the islands. An iron-ribbed, savage-toothed

186

shore the weather coast looked. Calm though the sea
was, the smooth, undulating swell, that only seemed
gently to rock the boat, shouldered the reefs with a
deep, earth-shaking boom and a rainbow-crested tur-
moil of foam. A smooth and slumbrous sea one would
have called it, from its offshore appearance, yet the
touch of that innocently-heaving swell had the force of
a million battering-rams. The softly-sliding billows
rolled caressingly against the rocks, and as they did so,
went up in a thunderous crash, spouting skywards in
columns of spray. Just to have touched the rocks would
have meant instant annihilation for the staunchest vessel
that ever floated. The unsinkable ship they try so hard
to build would have gone to pieces on them like an egg-
shell, as swiftly and completely as if she had hit the
Ildefonsos in a sou'-west gale.

Landing was out of the question, we dynamited
a few fish, and then pulled over to the nor'ard, coasting
the black precipices of 'Old Woman's Island' before
returning to the ship.

Just before Christmas the sloop brought word that
another ship, the *Annasona*, was coming to the Bay to
load. We were not particularly anxious to see her, for
the shore-gangs could hardly keep us supplied with
guano, and, since only two lighters were left, her
appearance would have lengthened our stay indefinitely. We watched the desert rim of ocean keenly
for sight of her white royals lifting, but day after day
passed, our holds grew gradually full, and still she did
not come.

187

We saw nothing of her; the Bay was still deserted when we left, and the *Annasona* evidently went else-where. Presumably she returned to the Colonies, for some time afterwards I read an account of her going ashore and becoming a total loss on Middleton Reef in the Tasman Sea.

Christmas Day we spent very quietly, doing no work and having duff for dinner. In the morning it was blazing hot, and we sat and smoked about the fore-deck, singing songs to Jimmy's accordion, and making the little penguins we had caught run races along the deck. Thorough little sportsmen they were, and waddled along manfully, the winner receiving a piece of fish as a prize. The old man sent the steward along with a couple of bottles of pisco, so we all had a drink apiece and wished each other a pleasant passage home. The wind freshened at mid-day, and during the afternoon blew a gale. We looked at the white-lipped seas out-side, and hoped for a spell of calm weather when we were loaded, for it would be impossible to get through the northern entrance to the Bay except with a steady, gentle breeze.

Next morning we returned to our unsavoury labours, and worked steadily away at the guano through an exceptionally windless week.

On New Year's Day we had the satisfaction of getting our last bag of guano aboard. We hoisted it up to the yard-arm and down again to the tune of 'Whisky Johnny,' with Gilroy sitting astride, energetically waving a flag. When the last shovelful had been trimmed down below, and the hatches were on, there was a welcome call of 'Splice the main brace,' and all hands laid aft to receive a tot of rum.

The decks were then washed, cargo-gear unrigged, hatches battened down, royals and stays'ls bent, and all hands turned in that night with joyful hearts in anticipation of soon being homeward-bound. We had to call in at Pisco to clear the ship and fill our water-tanks, and hoped that there we should hear our destination, and that it would be England.

Early next morning, the 2nd January, before the first glint of sun had touched the purple edge of the Carrasco Heights, we were roused out to take in our stern moorings. The gig was swung aboard, the anchors hove short, tops'ls loosed, and at eight o'clock we got under way. One of the seamen from the sloop, who had signed on for the passage home, clambered aboard and joined us for'ard, while our coasting-pilot, a Spanish half-breed, took up his station on the poop.

We rattled the anchor to the cat-head to the rousing strains of

> 'Blow, boys, blow,
> For Californ-i-o!
> For there's plenty of gold,
> So I've been told,
> On the Banks of the Sacramento!'

—and jubilant the chorus sounded, waking the echoes of that hill-bordered Bay. Everybody sang with a will, and when we sheeted home the tops'ls to the refrain of Shenandoah, 'the wide and rolling,' I was prepared to swear nobody had ever heard anything more wildly beautiful.

Before a gentle breeze, that just sufficed to give us steerage way, we moved slowly down the Bay. It was a lovely morning, the weather blue and unclouded, and all hands were in the highest spirits.

We did not leave the Bay by the passage by which we had entered it, but by an outlet at its northern end, called the Trujillana Channel, lying between Morro de Viejas and the main. Owing to the great number of isolated rocks, many of them just awash, that obstruct it, this latter passage is very narrow and intricate, with white water close aboard on either hand. However, we navigated it successfully, taking, as we did so, one last long look at our savage, sun-scorched furnace of a bay.

Once out in the open sea, we stood to the nor'ard before a freshening breeze, keeping close in to the land. The passage up the coast was very picturesque. The Andes hereabouts are of a prodigious height, and lie piled peak beyond peak in majestic confusion. We passed close in to Carretas Head, weathered the long promontory of Huacas Point, and by eight bells in the afternoon were abreast of the island of San Gallon, passing between it and the mainland.

The good trades were blowing freshly, and with slanting decks and straining canvas we sailed grandly through the narrow Gorge, the rocky ramparts of the main and the precipitous walls of the island looming high above on either hand. When night fell the wind again dropped to a gentle breeze, and a full moon sailed in splendour across a cloudless sky.

The weather conditions were so excellent that the old man, in consultation with the pilot, determined not to take the usual course between the Isla de la Ballesta and the Chinchas, but to steer in between the former island and the main continent.

This inshore passage is little more than a narrow gorge or cañon between towering walls of rock, and looks wild and romantic in the extreme, seen, as we saw it, from a sailing-ship's deck in the full splendour of a tropic moon. As we left the wide stretch of ocean, flooded with soft moonlight, behind, and steered straight into the dark, tortuous water-lane between the over-hanging mountains, it seemed almost possible to imagine we were leaving the confines of the cheerful earth and holding straight down into the gloomy jaws of the underworld.

For a few minutes we seemed lost and swallowed up among the roots of the mountains, gliding ghost-like past echoing walls of polished rock, with no sound save the lapping of the water and the faint reverberations of our voices. For a few minutes we slipped silently along, then, turning slightly to port, emerged suddenly into the sleeping waters of Pisco Bay. We stood in to the shore as near as was prudent, then let go our anchor, splitting the still night with the roar of twenty fathoms of cable. Sail was made fast, and, setting an anchor watch, we went below for the night.

Next morning we turned out and took stock of our surroundings. Before us lay the white walls and flat roofs of the little town, sleepy, ancient, and sun-baked, with the ruins of the old Spanish fort in the foreground, looking much the same as it must have done in 1586, when it was plundered by Sir Thomas Cavendish, the successor of Drake in the South Seas.

Close aboard us lay an ancient craft, flying the Peruvian flag, but in which Cavendish or Drake might have sailed. This venerable galleon was the only other vessel in port, and our eyes drifted to the distance, where, more remote than in the neighbourhood of Callao, rose the foothills of the never-ending Andes. I never looked away to that wild, mysterious, Sierra-bounded East, without a thrill of the heart and half-unconsciously murmuring the words 'I will lift up mine eyes unto the hills.'

Magnificent sight as they always are, the glittering Cordilleras do not display their full sublimity to gazers from the seaport towns. One cannot see the snow peaks from the lowlands of the Pacific. Far out to sea perhaps some isolated pinnacle may be descried soaring skywards in solitary purity — Aconcagua, for instance, is visible nearly three hundred miles in clear weather — but to see the whole range in its snow-crowned grandeur one must climb the coastal ranges, and then, across the naked deserts where the wind

'Sings to himself as he makes stride
Lonely and terrible on the Andean height'

is unfolded in majestic panorama the utmost bound of
the everlasting hills, icy, snow-crowned, and inaccessible – 'the watch-towers of the universe.'

As soon as we turned to, the bosun and four men were
sent away in the life-boat to bring off our water.
Several journeys they had to make, and horrible stuff
the water looked – green, stinking, and full of rubbish,
with an occasional small fish in it. The sight of it would
have turned a teetotaler to beer for the rest of his life,
but we, unfortunately, had no choice.

After breakfast, the Captain went ashore in the gig,
taking me with him to do any clerical work that might
be necessary.

A very few minutes sufficed to transact the ship's
business and, finding that he could not get his clearance
before night, the Captain suggested a trip to Ica. I, of
course, was delighted; and we had an interesting
journey and pleasant afternoon in that quaint old city,
catching a train and arriving back in Pisco just at dusk.
The outing made a delightful wind-up to our stay on
the coast, for the old man was a very hearty companion,
and the difference in our respective ranks was, for the
time being, overlooked.

Back in Pisco, the captain called in at the shipping
office to effect his clearance, sending me down to the
Mole to hail the boat. The business took longer than
he expected, and it was eight o'clock before he came
down with the agent. He took his place in the stern-
sheets and ordered us to 'Give way!' while the agent
called after him a cheery *Buen viaje, capitan!* We

guessed our orders had come, and fairly ripped the boat
out across the darkened waters of the bay to where her
lights showed the *Arethusa* was lying.

We drew the boat up alongside the gangway, and the
old man jumped out. Most of the men were standing
about by the rail in anticipation of hearing our destination, and as the captain stepped aboard he called loudly
to the mate, 'Get the boat aboard, Mr. Thomas, and
weigh anchor. Falmouth for orders!' We gave a hardly
suppressed cheer, and all hands set briskly about the task
of unmooring.

The gig was soon brought round under the davits,
hoisted aboard and lashed down in the chocks. Then
for'ard, and we manned the windlass, tramping round
the capstan to the tune of 'Good-bye, fare ye well,'
the unforgettable homeward-shanty.

As we hove round, straining at the bars, the 'clank-
clank' of the windlass-pawls was welcome music in our
ears. Our mud-hook was soon awash, the cat-fall
hooked on, and the anchor hoisted dripping to the cathead.

Then up aloft to throw the gaskets off and drop the
heavy folds of canvas. Tops'ls and t'gans'ls were
quickly sheeted home, and we began to move through
the water. A cast to starboard, and the open sea
lay ahead, with the lights of Pisco growing fainter
astern.

One after another, all hands working with a will,
our kites were spread to the gentle land-breeze. A
course was set, a hand sent to the look-out, and at eight

bells came the familiar order: 'Go below the port watch,' and we trooped off to our bunks, with smiling faces and contented hearts – Homeward bound!

CHAPTER TEN
Homeward Bound
*

HOMEWARD bound! there is magic in the words! To
the sailor sheeting home in some distant foreign port
what visions they evoke! Green fields and leafy lanes,
beloved faces and remembered happy haunts, village
bells on Sabbath evenings; all that's dear and hallowed,
all that's loved and longed for, rise in the heart and
gather to the eyes at the music of the magic 'Homeward
bound.'

> 'Rolling home, rolling home,
> Rolling home across the sea;
> Rolling home to Merry England,
> Rolling home, dear land, to thee.'

so runs the old sea song, and so we sang as we tramped
the deck under the gently-drawing canvas, with the
barque's head held down on the old trail, the long trail,
that leads round the stormy Horn.

One may love the sea and yet thrill to the sensation
of being homeward bound. To wanderers by profession
as well as inclination, who can say with Ancient Pistol
'the world's mine oyster,' the words are dear no less
than to the exile returning after long years in some
stony corner of the earth. It means, at least, a glimpse
of the green slopes of Falmouth, the excitement of
mooring in a busy city, a hearty welcome at home,
and a quiet spell before standing out to open sea again.

So with light hearts we bade a long farewell to the fast-receding shore-line of Peru. Yet I had grown to love that savage old West Coast with its everlasting rock and sun and bursting sea. A land shadowing with wings, a land of mighty mountains and Eden-like valleys, a land of mystery, romance and wonder – so seemed Peru to me. I visited it again and saw more of its fair cities and unsheltered bays; something even of the winding mule-trails in the recesses of its mountainous heart, but longer acquaintance only served to strengthen and confirm the fascination it exercised over me from the outset.

Even to-day the flavour of old-world romance lingers most persistently round that unregenerate strip of the old South Seas. An atmosphere of salt adventure haunts its lonely bays and steep sierras. Is it not the last refuge of those ever-diminishing sailing-ships, that are to-day all that remain of the great clipper fleets by whose aid Britain attained the very zenith of her maritime glory? Run off the rest of the oceans by the ever-increasing pressure of steam competition, white wings still linger there, and the bays and islands of the coast yet afford a precarious traffic to the last of their race.

Being homeward bound put everybody in good spirits. The fact that the long-desired day was still four months distant mattered not at all. We put a spring into every order we obeyed, and even our scanty sea-rations were accepted light-heartedly.

On the Coast we had been getting a plate of beans

and stewed beef — 'wet hash' sailors call it — for breakfast and supper. As soon as we were at sea again that
luxury stopped. The fish we had caught in Santa Rosa
now came in handy. The barrel in use stood under
the foc'sle head and we were not long in finding that
the contents didn't taste at all bad uncooked. As a
consequence they disappeared with astonishing rapidity.
We put one or two in our pockets every time we
passed the barrel and chewed them as we went about
our work.

There was much to be done as soon as we got fairly
out to sea. The first day we were employed in catting
and fishing the anchors. We lashed them down to the
foc'sle head, tautening each turn of the chain with a
vigorous 'chuck,' for they would have to stand many a
good battering ere we put them overside in the chops of
the Channel once more. The stream anchor was put
back in place and the mooring-chains chipped and
tarred before being flaked down in the forepeak. A
final trim was given to the cargo; the harbour gear
cleaned and oiled and put away for future use, and
soon all traces of our long stay in port had
disappeared.

Throughout our first day at sea the Andes lay like
a faint, grey wall all along the eastern horizon. Only
their sharp serrated edge, notching the sky-line, was
visible and with the coming of the shades of evening
that, too, faded out for good and all. The steady old
trades got hold of us and steering 'full and by' we
curtseyed leisurely to the south'ard.

A sheer delight it was to hand the wheel once more, with an eye turned aloft to watch for a tremor of the royal leaches, and no less pleasant to climb the foc'sle head for a two-hours' spell on the look-out. 'Mount Misery' we called the latter in bad weather, but in the region of the trades it was a delightful way of spending two peaceful hours of the starry night. Moreover, we were well manned now; with the coming of the sailor from the sloop there were five of us in a watch, only four of whom were employed at helm or look-out during each four hours of nightly duty. The odd man out was called the 'farmer,' and very pleasant it was to be 'farmer' on a warm tropic night with the good trades blowing. There was usually nothing to do but make oneself comfortable on a coil of rope and, lighting a pipe, look up at the stars and wonder 'who wouldn't sell a farm and come to sea?'

But no good thing lasts for ever, not even the Trades, and our days were employed in preparing for the passage of the stormy Horn. All too soon the golden days would melt out and merge into the mighty Westerlies and the rigours of the bitter South. We saw to it that we were not caught unprepared.

Everything aloft and alow was looked to; new rigging rove; the boats secured by means of extra lashings passed under the skids; mooring lines frapped round the coamings of the main hatch; capstan bars lashed to the stanchions athwart the ports, and life-lines rigged in readiness along the whole length of the maindeck. Every part of the ship was overhauled and prepared to withstand the onset of furious winds and grey-bearded seas.

In our leisure hours, too, we were not idle, as on the passage out.

All hands were at it, patching boots, mending clothes and re-coating oilskins. The constant chafe and soaking in salt water soon wears the protective glaze from an oilskin coat, and to renovate them we gave them two or three coatings of linseed oil, mixed in equal quantities of boiled and raw. The knightheads, the forestay, and the foc'sle head were soon a-flutter with oilskin-jackets and trousers hung out to dry. Each Saturday night we flocked aft and pretty well bought out what was left of the skipper's slop-chest.

In the half-deck we knew we were in for a wet and weary time, but we did what we could in the way of protecting the sky-light and making all snug. Our sea-chests we put in the lower bunks to be more out of the way of the water, but the doors were hopeless, and our best contrivance was to rig sacks as curtains just inside them, hanging from the deck above.

So for ten days or so we surged steadily to the south'ard. Almost every day we caught fish. Shoals of bonito crossed our track and on two or three occasions we made great hauls of them. A couple of hands were usually kept busy on the bowsprit in every watch below, while at the other end of the ship the old man baited and captured from morning till night. He caught any number of barracoutta and horse mackerel, a coryphene, too, and a couple of jew-fish, but the king of the

lot was an immense albicore that weighed seventy-
five pounds when gutted and cleaned. All these things
helped to eke out our scanty fare, and so did a flying-
fish or two that fell aboard and made a tasty supper
for its captor, if the steward could be prevailed upon
to fry it.

On our second evening out the old man called us aft
and told us it was time we started studying a little.
There was much to learn for the examination at the end
of our apprenticeship, and how did we propose to do it?
As a matter of fact we had given small thought to the
matter, but didn't tell him so and said we hoped to do
a lot of study when we got home at the end of the
voyage. The reply didn't seem very satisfactory to the
old man and he asked us a few questions about navigation, chart-work and the rule of the road that left us
stammering. In seamanship we got on better, but the
captain was not going to leave our education to chance
and took the matter into his own hands.

He told us to come along to the cabin every evening
from eight to ten and he would instruct us in the elements of our profession. This meant losing two hours of
our watch below every other night; but when it was our
turn on duty, the officers were told not to call us unless
there was occasion for it.

We passed out of the tropics on the 17th of the month
and the nights soon began to grow considerably colder.
We shifted sail a couple of days later and put up heavy
canvas in its place, for though the steady breeze still
held there came frequent gusts that betokened something

more behind. On the 23rd we sighted our first albatross
— sure sign that we were getting near the realm of the
West Wind — and the same day sighted the island of
Juan Fernandez.

This outpost of South America is a lofty verdurous
spot, with steep cliffs running sheer down into the sea
and a tumbled surface of woodland and savannah. It
is a favourite haunt of birds and seals, and its abundance
of vegetable and animal life, and advantages of wood and
water, made it a frequent calling place for navigators of
old after rounding the Horn. It became a regular port-
of-call for all vessels bound to and from the South Seas.
Drake was there and De Schouten; the Spaniards used it
at first, but it grew such a rendezvous for the privateers
that they gave up using it and tried to kill off the sheep
and goats, so that it might not afford succour to their
enemies. At the present-day the island is not nearly
such an important place as formerly; a few Chilenos live
there who can provide for the wants of storm-beaten
mariners, but for the most part all vessels, both steam
and sail, pass it by.

Close under the lee of Juan Fernandez lies 'Mas a
tierra,' a little barren rocky island in striking contrast
to its big green neighbour. It is so called to distinguish
it from 'Mas a fuera,' a much larger island, as green
and beautiful as Juan Fernandez, seventy leagues to the
westward — the two names signifying 'Nearer land' and
'Further off' respectively.

While passing Mas a tierra, but some miles distant,
the last of the little penguins we had caught in Inde-

pendencia Bay disappeared overboard. The little chap had been all alone for several days, his companions having one by one disappeared. He used to waddle disconsolately about the decks, or stand with his head poked out of a scupper-hole for hours at a time, watching the distant horizon. Apparently the sight of land proved too much for him and 'Hell or Melbourne,' thought he, 'here goes!' for he squeezed through and dropped into the sea. We watched him striking out for the land and I make no doubt he reached it with ease, for penguins are marvellous swimmers.

Two days after passing Juan Fernandez we had a heavy squall of rain. It was the first time we had experienced rain for close on five months and renewing its acquaintance was like running across an ancient enemy. It was accompanied by a longer and heavier puff of wind than usual, which forced us to let go the royal halyards and take in the sails. I scrambled down from the main drenched to the skin and shivering with cold, and as soon as I got below hunted up some warmer clothes than those I had been wearing.

That night while standing my two hours lookout on the foc'sle head my vigil was broken in upon by a startling apparition. It was a dark, starless night with a chill breeze moaning out of the east-south-east, and I paced back and fore across the break with my eyes searching the blackness ahead. As I walked briskly up and down I observed a pale light shape and collect itself out of the gloom, far distant but straight ahead of the vessel's track. I did not at once report it to the officer on watch, for it didn't look like a vessel's light but seemed more to resemble an atmospheric disturbance.

As it drew closer it resolved itself into a huge ball of pale, glimmering fire, seemingly dancing down on us on the crest of the waves. I didn't know what to make of it and sang out 'Light right ahead, sir!' to the mate on the poop. Mr. Thomas was evidently nonplussed too, for a moment afterwards our helm was put up and we fell off a couple of points.

The strange gleaming globe neared rapidly and from it came a sound of whirring and rustling, with the noise of raucous cries. Several of the men were looking out over the rail and 'big Mac' gave utterance to the general wonderment:

'What are ye?' said he, 'divil or man or baste?'

The uncanny object swept down on us, grazed our shoulder, and went swirling and gleaming by. As it did so the mystery was explained. It was a dead whale. Stripped of every scrap of skin, with its blubber exposed to the salt water, it glowed and sparkled all over with a shimmering phosphorescent light. The processes of decomposition had swelled it to a monstrous size, and its bulk lay on the water like a great bladder. Around it screamed and circled an uncounted multitude of birds, swooping, fighting, and tearing at it with their beaks. Every species of feathered inhabitant of the Southern Ocean, I should think, was represented; all squawking and uttering hoarse cries, as they hovered and picked greedily at the banquet provided. The grotesque apparition was gone in a few minutes, swallowed up in the gloom astern, with its luminous halo and noisy following. It made a very weird and not easily forgotten sight. What a ghost story it would have given rise to, I thought, if seen from a little distance by the daring but credulous mariners of old!

On the last day of the month we had our first introduction to bad weather – a good hard blow that gave us plenty of sail-drill and worked our back muscles up. The rain fell in a steady driving downpour and the amount of work aloft the mate found for us boys made us wish we were ducks. Oilskins are not much use aloft in heavy driving rain with the wind abaft the beam. 'Soul and body lashings,' as ropeyarn frappings round wrists, knees and waist are called, are powerless to prevent the back of one's jacket and the flap of one's souwester from being blown up, and the icy rain beating in and trickling down neck and back. Mr. Thomas had spent his life in a hard school, he was frankly contemptuous of our studies and he worked us up with vigour now that our evening lessons had temporarily come to an end. 'Kid-glove sailors' were his abomination and he saw to it that we came under no softening influence.

The wind had moderated somewhat by next morning, but there was a big lump of a sea still running. Soon after day-break we saw a large, full-rigged ship coming up astern. The mate crowded sail in the attempt to get away from her. But it was no use, she came on like a racehorse. She was abeam soon after breakfast, but too far distant for us to make out her name and, when eight bells were struck at noon, was hull down ahead. I should like to have known her name, for whoever she was, she was a flyer.

CHAPTER ELEVEN
Rounding the Horn
★

THERE is an old saying at sea that no one is a sailor until he has been round the Horn three times. Another – and highly undesirable – qualification is sometimes added, but that doesn't matter. The saying bears witness to the terrors of the Horn, and the grim old headland has not undeservedly won its reputation for endless gales and bitter weather. Of all the stormy waters of the globe, the passage of the Cape Horn 'greybeards' is the most to be dreaded. Sometimes, it is true, it is possible to catch the gaunt old sentinel asleep and slip round without a savage mishandling. On one occasion, with royals set, we passed close to the Ildefonsos, and, one day at noon, through the faint haze down the wind, saw the gaunt wraith of a rugged headland. It was old Cape Stiff itself, and before the weather woke up to a sense of neglected duty we were past the weird spikes of the Barnevelt Isles and had got comfortably through the Straits of Le Maire. But such opportunities are few and far between, and the Warden of the Southern Seas takes good care that few get past without paying him toll.

Nor did he forbear this time. The blow we had a week after passing Juan Fernandez was only the beginning of a succession of hard gales. All the time we held our way down the 'roaring forties,' gale followed gale from the south-west. A tremendous sea set up, and day and night the barque strained and laboured heavily. We showed nothing above the tops'ls, and the mainsail remained fast for days on end. Time and again we lowered the upper tops'ls and handed the sails, only to cast them loose and laboriously hoist the yards again, as soon as the hurricane weight of the wind gave promise of lifting a little.

During part of one day and night we lay hove-to under lower tops'ls. In this position, head-reaching to the wind, we took less water over in the waist but rode bows in to the running hills of seas. Solid green cliffs of water thundered down over the foc'sle head and surged aft in cataracts. But for the life-lines the main deck would have been impassable. As it was, several of the hands had very narrow escapes. Beckett was caught by a sea as he was running along the side of the deck-house and fetched up under the break of the poop with a cut on his head and a wrenched knee. At night the look-out was kept on top of the deck-house, no living man could have held his footing on the sea-swept foc'sle head.

Old Fox came near to losing his life through the scooping up of a sea. He and big Mac had been sent to re-stow the inner jib, which had broken adrift and was draggling over the knightheads. They laid out on the bowsprit, which at one moment was pointing to the sky and the next dug down to the depths. Of all the sails in a ship, the jib is the most dangerous to stow and has cost more lives than any other. The footropes hang directly over the sea, and the downward surge of the bows plunges them deep. The water hits the soles of one's boots with the impact of a solid body, and the utmost vigilance is necessary to save one from being knocked off by the blow, or sucked down when the water recedes to the following upward lunge of the head.

So it was in this case; a surging sea unshipped Fox, and he hung by one hand to the jackstay. Then up swept the head, and for a moment he dangled in air. The next he would have gone, if big Mac, who was on the other side of the bowsprit, had not reached over and grabbed him by the wrist. How Mac did it passes my understanding, but with one heave of his outstretched arm he lifted old Fox and swung him clear on top of the bowsprit. A more magnificent feat of strength, and done on the spur of the moment, I never saw. But it passed practically unnoticed; the sail was made fast, and the two scrambled in again and, watching their chance, regained the main deck. Such averted mishaps are too common at sea to call for much comment or remark.

On the third of February we crossed the fiftieth parallel, and, contrary to expectation, the wind moderated a little. A huge quartering sea still stormed at us, but the wind no longer felt like a wall, and the old man seized the opportunity to shake out the t'gans'ls.

Our good luck was not to last for long. The wind soon began to rise in squalls of increasing violence. It was my wheel from ten to twelve that morning, and just on the stroke of six bells a heavier blast than usual struck the ship. Over she lay, a wail of warning rising from tense steel and straining wire – over and over until the sheerpoles were level with the water. The wind's note rose almost to a scream, and every line aloft was rigid to the strain.

The captain shouted an order to let fly the t'gallant halyards, and jumped to the helm to heave up on the spokes. Even as we hove there was a sharp report for'ard, followed by a terrific roar. I saw the rush of a falling spar and a wild confusion of rope and canvas.

For a moment I thought the fore-topmast had gone. It was not that, however, but the tops'l tye that had parted. The yard came down with a crash, both t'gallant sheets parted and the t'gans'l itself soared up and was gone.

The heavy yard hit the top-mast cap an awful blow, splitting it in two and smashing the parral. The sail went to shreds, and in a moment the barque was a howling wreck for'ard, with the foremast a tangle of spars, ropes and fluttering canvas.

The devil's own job it meant for us. In our disabled state, with a heavy sea running, it was dangerous work to lay aloft at all. The tops'l yard was grinding heavily, with a splinter of steel at every roll. The safety of the lower yard was precarious, while, with the cap split and its backstays useless, the whole top-mast stood loose. The old man was the first aloft, and with the second mate and two or three of the men temporarily lashed the crazy contraption to the lower mast.

It took some time to do that, but this was only the beginning of it. We were helpless without our fore-mast, and the 'hurrah's nest' aloft had to be made fit to carry sail again.

For three days all hands, with brief respites for meals, laboured at it. A wire preventer lashing was put on under the cap and the backstays shackled to it. Both tops'l yards were jury-rigged. The parral of the upper was repaired and the crane of the lower supported by a truss over the topmast head, that required constant attention every time the upper yard was touched.

What with the tumbling about of the ship, the bitter cold, the endless hours aloft and the merciless work with stubborn wire and steel, all hands were pretty well dead-beat when the job was finished. All the rest we got was in our watch below at night; cuts and bruises were the lot of everyone, but we were driven on by dire necessity, and the old man himself led the work.

When at last the damage was repaired and we were able to make sail on the foremast again, never a saint knows the road to heaven as we knew the way up our foremast rigging, and every turn and bolt and seizing around the doublings of the mast.

A few days later another mishap occurred. The bob-stay — the stout chain that holds down the bowsprit and is shackled to a bolt in the cut-water — carried away. In the sea that was then running it was hopeless to think of repairing the actual stay itself, and we had to rig preventer stays on either side of the bowsprit, setting them up through the hawsepipes. A wet job it was, particularly when we came to knock out the hawse-plugs and bowse the preventers tight. At every plunge of the ship cascades of water surged through the pipes, but it was child's play after our previous experience.

The hard swearing of the almost-drowned men at the watch-tackles even changed to a chuckle for a moment, as the second mate, bending down to hear an order from above, caught a sea full in the ear and was washed up against the windlass, spluttering anathemas.

But all the time, endless gale though the wind was, at least it was fair, and every day we drew nearer to the pitch of the Horn.

One morning at day-break we sighted a small barque, outward bound, and were able to thank our lucky stars we weren't on board her. She was plunging close-hauled into the fierce squalls, with only her lower tops'ls set. Her crowd must have been having a lively time, for she was digging into the seas as though she were trying to scoop the South Pacific up and throw it over her shoulder. We ran past her very quickly. She was port-painted, and, I should think, by the cut of her, British. The weather was too bad for us to speak, and she was soon swallowed up in the mist and spray astern.

The day after passing the port-painted barque, we sighted the Diego Ramirez Islands — the Dagger Rammer E's, sailors call them. Beckett had the honour of sighting them.

When setting the lookout the night before, we had been ordered to keep a sharp watch for land. It was Beckett's look-out from six to eight next morning, and just at eight bells, as a haggard grey dawn was lightening over a wild sea and ragged sky, he saw a cluster of lofty rocks in a leaping white sea, broad out on the port bow. He hailed the poop with a lusty 'Land-ho!' and out-

stretched arm, turning all eyes to the iron-bound fragments of land. The old man came up on the poop, and
we soon learnt it was the Islands right enough.

As day broadened we saw them more distinctly.
Unspeakably wild and savage they looked. We saw the
loom of a bigger island behind, and they must have been
surrounded by sunken reefs and ledges, for everywhere
the sea was breaking white, seething and spouting in a
wild field of foam. Gilroy and I struggled up on to the
foc'sle head to get a better view of them, and, as we did
so, the former's remark condensed pages of description:

'Good Lord!' he said, 'what a hell of a place!'

Indirectly, and without unnecessary verbiage, that
just about sums up the Diego Ramirez.

With the islands astern our course lay pretty well due
East, and for some days we had terrific weather. The
cargo didn't trim well; the foremast was a constant
source of anxiety, and the barque laboured very heavily,
complaining in every plate and beam of her.

Looking back on it, the week or so that followed
looms up like a nightmare in my memory. It seemed a
period of perpetual darkness and the fierce buffeting of
one incessant gale. We appeared to make no headway.
Day and night was an endless round of work in icy
water on the swirling decks, or aloft in driving squalls of
hail and sleet. Ice froze on the yards and the ropes were
coated stiff with freezing spray. Life under such conditions was nothing but a combination of cold, darkness
and misery.

Our rotten old half-deck was two feet under water

the whole time. Every stitch of bedding and clothing we possessed was wet through; not a dry shirt had one of us to his name. We vowed it would have been cruelty to house a pig in such a place. And almost like pigs we lived. Unwashed, unshaven, never even removing our sea-boots, we wolfed our scanty meals anyhow and turned in all-standing.

Little, we growled, did our folks at home know what a sea-life was like. But just wait till we got home, we'd tell them; we'd prevent anybody of our acquaintance from ever coming to sea. And then we would laugh savagely at the idea of passengers on a liner thinking they'd seen the sea, and describing themselves, if they escaped a bout of sea-sickness, as 'good sailors.' 'Good sailors,' egad! Anyone who'd go to sea for pleasure would go to hell for a pastime. Curse the sea! And curse the ship! Why shouldn't we curse ? All very well for parsons to talk, but put a bishop on that blasted tops'l yard with finger split to the bone, and let him wrestle in the darkness and icy wind with the board-like canvas and he'd say other things beside his prayers.

And so on, with much more to the same effect. We meant it, every word of it, at the time, but it only lasted as long as the weather. The sea is a changeable mistress; at times a tender nurse, at times a smiling Circe, and yet again a fiend-hearted, slaughter-breathing fury. Her servitors vary with her moods, and before judgment be passed upon them it were well that the judge had full and ripe experience. The loneliness of the life, the hardness of much of it, the sudden changes from storm-

beaten penury to the gilded pleasures of sailor-town,
make for the greater virtues and the greater vices. The
meaner of the latter find no place on board ship. Whatever the sea-farer does, he does largely. His ways are
not the ways of sheltered folk ashore, nor are his morals
theirs, nor his creeds. Talking of a shipmate behind his
back is a far blacker vice than swearing, and selfishness
more to be condemned than insobriety. And if accepted
creeds sound thin at sea and catechisms of small importance, the sailor is at heart no less religious a man – and
no worse a one.

In spite of all our discomforts, and they were accen-
tuated by the boils and sea-sores that covered neck and
wrists through the constant chafe of salt-soddened
clothing, we yet had something to be thankful for. The
great Westerlies held true to their name and we had no
headwinds to contend against. Our progress was slow,
but not through time lost beating into the teeth of the
wind. The furious gale kept ever on the quarter. It
rose and rose in successive blasts until the culminating
blow leapt at us like the breath of the hurricane, yet
never hauled before the beam. And the sea was worthy
of such a wind. It was a wonderful, a magnificent sight.
From horizon to horizon the rollers stretched, undulating hills with snowy crests. In those latitudes there's
never a ridge of land to check their world-encircling
sweep, and the majesty of ocean is nowhere so pre-
eminent.

Small wonder that the apex of the South American
continent is slowly being worn away before the unceas-

ing siege and onslaught of the sea. Those mighty rollers
have worn the barrier thin. The western sea-board has
been fretted into a million reefs and islands and isolated
rocks. Long tongues of sea run up into the land. The
lofty, precipitous coasts have been hacked and hewn into
deep fiords, and archipelagoes of multitudinous islands
lie off their seaward outlets.

Some little protection the islands, smoking white in
the spume of the sea, afford to the encroachments of
ocean. Iron fragments as they are, they are making a
stern fight of it and all, save the weather-outliers, those
grim and naked tusks that take the first brunt of the sea,
call vegetation to their aid and in every sheltered nook
and cranny are clothed with trees, that bind their soil
together and enable them to hang on.

It is a Titanic combat, this elemental warfare of
tempest-winged water and stubborn land, but the former
is winning. Countless ages it may take, none the less
the victory is certain.

To see the black, forbidding bastions of land when,
for a breathing space, the winds are hushed in the halls of
heaven, one would say that neither time nor lightning-
stroke could prevail against them. To see those same
spectre-rocks in the boil of a white-lipped hurricane
makes one realize the ultimate outcome of the conflict
is sure. If the trumpet of Michael does not sound first,
one distant day the waves of the Southern Ocean will
ring the Roaring Forties round, with no continent
athwart their path and only the kernels of islands
a-smother in their wake.

ROUNDING THE HORN

To descend from the sublime to the insignificant —
those same mighty waters were very nearly the death
of us.

We were off the pitch of the Horn on the tenth of the
month, and the weight of the wind rose to hurricane
force. All through the previous night we had run with
only lower tops'ls set; the coming of the sullen dawn
brought no cessation in the wind, which still rose and
rose in blinding, unfaceable hail-squalls. In the forenoon
the fore tops'l was taken in and the only canvas left
spread was the slender strip of the main lower tops'l.
We goose-winged that soon after, taking turn after turn
of ratlin-line round the bunt, to keep all firm, and leaving only the fins of the clews yet spread to the gale.
Few square feet of canvas as were left, it was all the
barque could stagger under. Bare poles would have
been enough.

Strange it was to look aloft and see the reeling spars
gleam yellow and naked against the leaden sky. In
place of the sheeted tiers we were accustomed to, nothing but the wet gleam of painted steel and the tensely-
drawn lines of unquivering cordage. Fore and aft no
stitch of canvas except those rigid, straining triangles
at the main. To the hail-whipped fury of the squalls it
seemed that bare steel and wire itself could hardly stand.
The sea was a moving mountain-range of water; a
broken expanse of green and white, running true as
though laid by line, and with unfaltering swing. Half
the world seemed to heave up as we shouldered some
league-long crest; not a cable's length could we see

217

when the squalls swooped down. In the height of the blasts the crests of the rollers were blown headlong in a flurry of spindrift that hung like a mist.

There was no chance to wear or heave-to. To have brought her broadside on, if only for a moment, to those mighty following seas would have meant instant destruction. Like the Alexandrian sail of old, the ship 'was caught, and could not bear up into the, wind.' There was no alternative but to let her drive.

Only our best steersmen were allowed at the wheel. Another man was stationed to leeward to bear a hand with the spokes. The old man never moved far away from the helm. The remainder of us were gathered on the fore part of the poop. There was nothing we could do while the gear held. The maindeck was a seething whirlpool swept by endless tons of green water, that made a dash for'ard a matter of life and death.

There was no galley fire, and the steward remained perforce in his pantry under the poop. He managed to boil some water on the mate's oil-stove and we had a drink of tea apiece at mid-day, with a sea-biscuit and a handful of bully-beef.

So the day wore on. Drenched to the skin and strained almost to breaking point through standing up against that living wall of wind, I thought the end of the world had come – and come it nearly did for us.

It was just after four bells in the afternoon watch, following a shrieking squall that exceeded all that had gone before, and while the barque was still wallowing like a half-tide rock, that we saw a monstrous, foam-crested breaker rolling up astern.

We had seen many before; this was but a giant among giants, coming on with unopposable stride. It was the behaviour of the hardly-pressed barque that troubled us. She lay like a gladiator, sore-stricken and fainting, care-less of the clamour around and the uplifted sword of an exultant foe. Buried deep under a weight of water, there was no life in her.

Higher and higher the comber rose, with a toppling, concave crest, swiftly overtaking the ship.

We doubted if the staggering hull would ever rise to it and there was little need for the old man's hoarse shout: 'Hang on all!' We sprang for stanchion and backstay and clung desperately to them. The towering grey-beard swept down on the ship, came up with her, and was met by no answering rise.

High above the taffrail — forty or fifty feet — it loomed, and the next moment it fell.

The fall of the firmament from above could not have been more terrible. Six feet above the poop deck we were buried under a black weight of water. For the space of a few seconds we knew not if the barque still floated or was being forced down to the depths inexor-ably. Through instinct more than exercise of will I hung on, with the strangle-hold of a nightmare upon me and the deadly thunder of water in my ears.

I felt my shoulders were being wrenched out as demon-fingers plucked at me; then the weight of the avalanche lifted, and I knew the blessed feel of light and freedom again. It cannot have lasted for more than the space of a few seconds, but it sufficed me to learn the meaning of that word that in eternity a thousand years are but as a moment.

The bosun and I had jumped for the port mizzen rigging and had been clinging to the topmast backstay. As the water passed we looked up. The great roller that had pooped us swept for'ard and buried the ship deep under a green swirl of water. Even as we looked, two walls of water rushed in over the submerged bulwarks and collided down the length of the ship. The hull settled and felt dead beneath our feet.

'My God, she's gone!' said the bosun.

I glanced at the fore t'gallant yard, motionless against the sky. It was the last thing I ever expected to see. Nothing of the ship was visible, save the deck of the foc'sle head, like a lonely rock. Another bucket of water would have done for us.

For a few seconds we lay, as it were, stricken and a-swoon. Another white-lipped monster was rolling up astern, but before it reached us the gallant old barque seemed to make a mighty effort. She quivered and laboured heavily up, throwing the water from her main deck and lifting her streaming bows. As the roller swung down on us, her stern rose slowly to it, and it surged on and under, lifting the barque on its shoulders, spouting cataracts from every port.

The worst was over; we had come through that and were ready to face fresh onslaughts with confidence.

But what havoc it had played with us! As I looked

220

aft I saw big Mac at the wheel and the old man, bareheaded, at his elbow. The wheel-box had utterly gone and the well-oiled steel couplings stood bare to the spray. The binnacle still stood, but the cabin skylight had gone to matchwood. Tons and tons of water had fallen below, flooding the cabins and filling them with a litter of wood and splintered glass. The top of the lazarette was smashed in, and the fragments of the scuttle held only by a broken hinge. All the poop gratings, the covering board and the weather cloth had disappeared utterly. The poop was swept bare. What had happened on the main deck we couldn't see, but the flying bridge was smashed and the boat lashed on top of the house – a full seven feet above the deck – was stove in and lay, a dejected raffle of boards and broken edges, held together only by the lashings in which it was swathed.

Nor had the men escaped. All hands were there, but more or less battered. Several of them were bleeding; the second mate was propping John Neilsen, white-faced and barely-conscious, up in the companion. The old mate lay under the mizzen rigging with his foot doubled up in a tangle of ropes, unable to rise. The steward, who was below in the cabin, was nearly drowned. There was no escape for the water up to the height of the pantry flap and it lay three feet deep in the saloon and cabin, swishing about with a wreckage of wood on top and a reef of broken glass underneath. He was all night bailing it out.

All through that shrieking afternoon, with never a

jot of abatement in wind or sea, we ran blindly on our way, two men at the helm and the old man, broad-shouldered and bare-headed, standing before them, conning the ship. The binnacle was useless, broken for all we knew, for the Flinders-bars had gone with the skylight. Eye alone had to guide the ship now. Night came, and still the old man stood there, and next day broke and he had not once moved away. He rarely spoke, but with eyes ranging to port, to starboard, and aloft, directed the steering with motions of hand and arm. The navigators among us were fond of criticizing the old man, but that night he silenced criticism, as far as bad weather was concerned, once and for all. We should have been in bad plight but for his skill and endurance.

Fortunately our good main tops'l stood. We carried nothing away aloft, though several times we had to lay out on the yards and secure a sail that was in danger of being blown adrift. For the rest, all hands could only shelter on the poop – cold, hungry, and heavy-eyed, waiting for a break to lighten to wind'ard.

The fury of the gale moderated a little on the following afternoon and we were able to repair the broken skylight and restore the water-logged cabins to some semblance of order. The old mate was put into his bunk, badly wrenched and strained, but Neilsen hung on to his duty, though his neck and shoulders went black, through the violence of the blow he had received.

The wind still lessened, and at nightfall we shook out both lower tops'ls and were able to bring the ship to her course. On the following morning it had dropped to a fresh gale and we set the upper tops'ls. The old mate gamely struggled on to the poop and directed the work from there. Tommy coaxed a fire into the galley-stove and we had a welcome cup of coffee. Chips was kept hard at work boarding up the skylight, and the captain spent a long time adjusting the compass as best he could.

This gale blew itself out, but hard, stormy weather followed us for some time longer. Nothing so bad as our experience of pooping occurred again though, and under double tops'ls we ran wallowing to the north-east.

As it was, the Horn had given us a rough baptism. We could not well have been nearer foundering than we were when that roller broke on us. Of the four times that I rounded the Horn in sail this was far and away the worst. I believe that it was an exceptionally stormy year, even for old Cape Stiff. A number of vessels had unusually rough handlings down there about that time. The experience of one vessel – the *Celtic Monarch* – was so bad as to find a permanent place in *The South Atlantic Directory*. And little enough as it sounds on paper, our own experiences were worse in the encounter than in the recapitulation.

It is impossible to describe the might and majesty of a Cape Horn sea. Words are not capable of such a thing. One is tempted to pile adjective on adjective and revel in terms that from the mouth of roaring Typhon dropped would seem hyperboles. It's all useless. Those who have not seen it cannot imagine it, and those who have need no shake of memory's wing. Seen from the deck of a little barque of a thousand tons or so, such a sea is the high-watermark of elemental majesty.

Terrific though it is, it inspires no fear. The wonder and the majesty of it blot out all that. A winter's gale on a lee shore is a nerve-racking experience. Not so a Cape Horn sea. One merely triumphs in the exhibition of such stupendous power and sublimity. Death itself would be a small thing in such surroundings. One could pass untroubled through such mighty portals to

'the idle tides of eternity.'

By the 14th of the month the barque had been restored to a more normal appearance and we were pursuing our usual 'hard weather' routine when we passed Beauchene Island.

Beauchene is an outlier of the Falklands, a rugged, uninhabited islet, rock-fringed and flat-topped, a quite insignificant member of the famous group. The two main islands are East and West Falkland – the former of which we sighted soon afterwards.

The handling we received farther south did not necessitate our putting into Port Stanley for repairs, as many a Cape-Horner is compelled to do. On a subsequent voyage the harbour was a haven of refuge to us, and the first object I saw on entering it was a small old-fashioned paddle-wheel tug, called the *Flying Childers*, that many years previous had been a familiar sight, during holidays at home, towing fishing boats in and out of Yarmouth. Now she was ending her days, a world's width away, playing the Good Samaritan to limping Cape-Horners.

Every night at eight bells as the watches were changed and wheel and look-out relieved, order was given to keep a sharp look-out for ice. Often at this season of the year the seas south and east of the Falklands are covered by floating ice, but we only saw one solitary berg. It was the day after passing the Falklands and we were forewarned of its proximity by a sudden drop in the temperature. A fine sight it presented, gleaming green and blue and white and cruising statelily along. Bergs are ugly neighbours and we kept some distance from it, but we reckoned its length at a quarter of a mile and its height at perhaps a couple of hundred feet. To blunder into such a thing at night would be sudden death, but, serious danger to navigation as icebergs are, they fortunately give off a white blink that often serves to avert disaster, even on the darkest night. On the other hand, a strong current usually runs along the sides, setting in against the steel-like walls of ice. In heavy weather they are more menacing still, for the great seas do not bodily lift the berg but burst all over it, as though it were land firmly moored to the bed of the ocean. It is a matter for congratulation that they do not seem to drift in close to the Horn, though they are met with much farther north, both east and west of it.

With the sun shining, an iceberg is a glorious sight, gleaming and sparkling with a hundred lovely hues, but even so, I never yet met the seaman who would not willingly forego the pleasure of seeing, to be free of the risk of touching, them.

Notwithstanding the rigours of the weather, with

the Falklands astern we commenced our long task of cleaning the ship in readiness for our arrival home. Painting and the niceties of decoration were out of the question so far south, but the opportunity was taken to holystone the decks and scrub the brightwork clean. Down on our knees we went at it, scrubbing for four hours at a stretch, getting pleasantly wet through, as little dollops of water dropped in on us, and rubbing most of the skin off the tips of our fingers into the bargain.

So we slanted north, encountering wet, squally weather, across the breadth of the roaring forties. Every night the Magellan clouds grew fainter and fainter astern, 'and the old lost stars wheeled back.'

We met with no more ice, and the weather grew gradually warmer. With the fortieth parallel astern we thought we had done with bad weather, but were to get one more blow by way of a final blessing.

There's a rhyme at sea to the effect that

'When Cape Horn you safely pass
Then look out for Hatteras'

meaning that these two notorious storm-centres will tax the seamen's skill, but that the latitudes between them do not offer much danger to navigation. The saying is sound enough in a general way, unless one blunders into a West Indian hurricane, when it may be said to leak a bit. The estuary of the Plate, however, can produce its fair share of dirty weather, though compared to the terrors of the Horn it is an exhibition of mere childish petulance.

ROUNDING THE HORN

When we were somewhere off the mouth of that river we ran into a 'pampero.' These are hot violent winds blowing directly off the vast pampas of South America – hence their name. They are not of long duration, but of terrific intensity and almost unheralded.

This one caught us without a moment's warning. It was about one bell in the second dog-watch; the ship was under t'gans'ls, and the bosun and I were yarning under the side of the house, when a black, funnel-shaped cloud raced up out of the nor'west with tremendous speed.

The mate saw it coming. 'Let go your t'gallant halyards,' he shouted, hobbling aft to help the helmsman. The words were not out of his mouth before the squall was upon us. There was a roar like a million bricks falling, and a sensation as though we were in the middle of them. I let fly the fore t'gallant halyards and Stedman scrambled over to the main on the opposite side of the deck. The barque was lying over at such an angle that he had to dig his fingers into the seams between the planks to claw his way to wind'ard. Even when the halyards were let go, the masts lay over so much that the yards would not budge, but stuck at the masthead.

All hands tumbled out on deck and, manning the clew-lines, hauled the yards down by main force. Stays'l halyards were let fly and hauled down and the barque righted to a more normal angle – for a few minutes her sides had been nearer the horizontal than her decks.

The wind rushed at us like a witch on a broomstick, the air grew black as pitch, and suddenly a rattling volley of thunder cracked and banged above our heads. Without hint or warning it exploded in an almighty crash. In an instant the lightning blazed out everywhere, zig-zagging across the solid vault on every side and seemingly stabbing through and through the ship. For some minutes it was hell with the lid off. Then, with a hiss and a roar, down came the rain!

We worked blindly and frantically at the ropes in our efforts quickly to snug her down. The devil's tattoo of the wind, the pitchy darkness and the deafening thunder-roll overhead made concerted effort impossible and we worked as best we could individually, obeying orders when we heard them.

'Here,' would come an urgent voice, blown whistling past one's ears, 'bear a hand . . . this blasted clew-line!' A shape or two surged forward and a rope was shaken violently.

'You've . . . got the buntline. This . . . here!' A drenched figure bumped violently against you, a wet rope flicked your face and you grabbed and hauled madly at it.

Stumbling, cursing, half our efforts aimlessly spent, the t'gans'ls were clewed up, the stays'ls hauled down, while the fury of the squall lessened, the thunder dulled and the cataracts of rain settled down into a steady downpour.

It was still blowing great guns as we lay aloft to furl the bellowing canvas and restore the stampeding ship

to some sort of peace and order. We snugged her down
to tops'ls before the order was given to relieve the wheel
and the watch sent below.

All that night it blew hard, and a busy time the
watch had till daylight broke and the pampero sullenly
blew itself out. Then we made sail and resumed the
tenor of our day's work. In the brief log I always kept
on my wanderings I find this encounter with a pampero
summarized in the entry, 'Incessant thunder and
lightning; black as soot, and rain like water-spouts.
Damn bad night!'

CHAPTER TWELVE
'Beating up from Southerly'

*

CLIMBING the hill for home is always the grand opportunity for cleaning ship. A serious business this of furbishing up is. All hands are busily occupied at it from the roaring forties to the Western Isles. A preliminary scrub is often attempted as soon as the Falklands are astern, and final pats of paint and fancy seizings are being put on till the vessel is on the edge of soundings.

There is much to do. Up aloft the masts and spars receive two or three coats of paint; every scrap of the standing rigging is oiled or tarred down, ratlines are set up, blocks overhauled and fresh seizings put on. On deck there is the bright-work to be scrubbed clean, houses and bulwarks to be painted, planks to be holystoned, brass polished and boats overhauled and redecorated. The lockers under the foc'sle head, the galley, the half-deck and the men's quarters – all have to be painted. There is besides much delicate painting of the ship's name, in cunning scroll work, on buckets, lifebuoys, boats and wheelbox.

All this is in addition to the regular work of the ship and entails many a watch below being spent on deck. The greater part of it is pleasant enough work, though the apprentices, as usual, come off second best. Being esteemed nimble fellows, all the spidery work near heaven falls to their share. Gilroy, Jimmy and I took a

230

truck each to work down from, and I unluckily got the
mizzen. Being barque-rigged, there was less work to
do than on either the main or fore, but it was far more
awkward and un-get-at-able. There were no ratlines
above the lower-mast cap and one had to shin up a backstay the sixty feet to the mast-head. With a pot of paint
dangling at one's belt, it necessitated careful climbing,
for heaven help the man who spilt a drop on the newly-
scrubbed decks.

The gaff was even worse; it ran out from the mast at
an angle of nearly 45 degrees and there wasn't a single
thing to hold on to. One crawled out to the end,
then slipped and slithered back, painting as one came.
The very real danger of falling was forgotten in the
acute fear of spilling paint on the resplendent poop
beneath.

Tarring down, too, is dirty work, for which we
donned our oldest clothes. But these disagreeable items
were offset by long, pleasant hours aloft, spent in serving,
parcelling and seizing, or engaged with a pot of paint
somewhere about the decks.

The worst job of all – holystoning – we were pretty
well through with. We had done it in the wet weather
further south. I don't know a more unpleasant job
than that of holystoning, particularly if the weather be
cold and rainy, with little seas slopping over the weather
rail and catching one helpless and unawares. Down on
hands and knees for four hours at a stretch, one
soon gets wet through and acquires an aching back
through keeping such an uncomfortable posture on

231

the slanting planks. The deck seems an interminable expanse, and most wearisome the mate's reiterated 'See you don't leave any holidays' – 'holidays' being scamped work.

The blow we had off the Plate was our last touch of bad weather. Fresh variable winds carried us up until we got hold of the Trades, and the business of painting, scrubbing and sailorizing went on apace. The captain overhauled his fishing-lines again and made several fine catches of barracoutta and bonito.

Though we had experienced no more than our fair share of head winds, the ship's progress was very slow and our hopes of making a good passage grew every_day more remote. Our stock of water, we found, was running low and all hands were put on three-quarter allowance. The ship's bottom was evidently very foul and Chips was turned to, making a rough-edged wooden grating with which to scrub it when we got into the region of light winds and doldrums. The carpenter had before this finished repairing the poop skylight and made a seamanlike job of it, but there were no means of replacing the glass, and the cabin was a draughty place when the canvas skylight-cover was not on.

As soon as ever the weather became moderate we renewed our evening studies in the cabin. We had made little models to teach us the rule of the road and semaphores for signalling and quite looked forward to our nightly two hours. The old man dropped in on us at intervals and unravelled knotty points. He made us work up the noon sights every day by way of practice.

I worked a Sumner, or position by double altitude, in the South Atlantic and it pleased the skipper so much that he gave me the chart on which I plotted it out to keep as a memento.

Another of our duties was to write up the log-slate hanging in the companion and take it to the mate for his approval. Two logs are kept on board ship – the official log in possession of the captain, wherein are entered details of deaths, injuries, advances to the crew, collisions and the like; and the navigation log, written up by the Chief Officer, and containing particulars as to position, wind, course, speed, weather and details of work done. It was this latter that we had to write entries for and we cultivated assiduously the peculiar style of English required for such work. Poetical disquisitions were not encouraged and one of us got a terrible wigging for putting in a remark about 'the rolling deep.'

However, that's by the way. It was sheer kindness on the captain's part to take such an interest in the education of his apprentices. We hardly appreciated it at the time at its true value. Since then I have learnt that the commanders who bestowed such attention on their apprentices were few and far between. The owners rarely interested themselves in the matter; it was left entirely to the captains and we had much cause to be grateful to Captain West.

Our commander's masterful manner we had long grown accustomed to. The terrific vehemence of his method in command never altered, but off duty his

grimness relaxed and he was almost fatherly in our
evening yarns and lessons in the cabin. Deep down
in him there were unsuspected depths of kindliness,
but the captain of a sailing-ship is a lonely and auto-
cratic individual and rarely wears his heart upon his
sleeve.

His very masterfulness, I soon came to think, was
one of his greatest virtues. He was always the absolute
ruler on his own ship, and, being the captain, this was as
it should be. I have since sailed with men of many
characters — severe, soft and nondescript — and, having
done so, am all for the man who is master of his own
vessel, even if he is a bit of a Tartar, rather than the
milk-and-watery individual who lets his crew take
charge. In the first place you have only one master;
in the latter you have a dozen.

I don't advocate American methods, though I once
saw Captain West pick up an insubordinate seaman by
his thigh and throat and drop him over the poop rail
to the main deck ten feet below, and it had a wonderfully pacifying effect. Discipline can usually be main-
tained without 'belaying-pin soup,' and the days of
Bully Waterman and the Yankee hell-ships are
fortunately over and done with. But the pendulum
now seems swinging the other way and a reversion
to windjammer ideas of discipline would be a healthy
change.

But we had no need of such methods on this voyage.
We had a fine crowd for'ard that required no 'hazing'
to make them do their duty and, on the homeward

passage in particular, were a happy ship. Our long interrupted sing-songs were resumed with the coming of the south east trades, and though Paddy was no longer with us, the bosun made no bad substitute.

Sing-songs were almost our only amusement. Reading was impossible, for the very good reason that we had no books left. The few that had survived the West Coast had succumbed to the rigours of the Horn and been dumped, a sodden pulp, overboard. My battered old Shakespeare was the only book left in the half-deck and I hung on to that with grim solicitude. It was the Globe edition and I often blessed the serviceable paper and neat print – less good workmanship would never have stood so much salt water. We often read scraps out loud, and, on one occasion, when the bosum came in I fired off the first scene of the *Tempest* at him. He was immensely taken with it, but would hardly believe it was Shakespeare at all. However, he knew what 'bringing a ship to try' was, which was more than I did at the time or, I dare swear, a good many others who have read the play.

It was not till the 5th of March that we entered the tropics. The Trades blew steadily but gently and our speed rarely exceeded eight knots. Though our progress was slow, it was under ideal conditions – sparkling sunny days and warm, lazy, star-spangled nights.

The old man was greatly concerned about the steering-compass, that had been damaged when we were pooped off the Horn. Now that the Trades were blowing and we were steering 'by the wind' he spent long hours trying to adjust it, in conformity with the standard instrument on the flying-bridge. He was successful after a time, and before we reached the Line we were able to rely once more upon our main aid to navigation.

A ship's compass is so indispensable that its necessity is apt to be overlooked, but a mishap to it soon brings out its pre-eminent importance. Whether in steam or sail it is the last instrument one could dispense with. The compass is one of those simple inventions that have shaped the course of history. Without it, we should still — to use an Irishism — be coasting on our deep-water passages, clawing our way from cape to cape like an Arab dhow.

Another indispensability of ocean navigation now began to give us anxiety — our fresh water, to wit. In spite of the short allowance we had been on for weeks our supply continued to dwindle at an alarming rate and Chips, who sounded the tanks every morning, reported a leak in one of them. It was impossible to get at the damaged place and repair it. All we could do was still further to economize and hope for a down-pour of rain in the doldrums, which would enable us to replenish our stock.

Besides the shortage of water, tobacco was another commodity of which we were running short. Tobacco never tastes so good as in the salt air on ship-board, and is almost a necessity of existence in a hard-case 'lime-juicer.' A pipe of tobacco has often to serve in place of

a meal and, rammed well down and glowing red in the
bowl of a short clay, makes the cheeriest of companions
in the long night watches. Anyhow, we felt the loss of
it very keenly; the last few plugs changed hands at
fancy prices, and when they, too, were all gone our
search for substitutes was exhaustive. Ropeyarns,
whether manilla or hemp, ravelled out, were tried and
found very hot and heady; tea-leaves had their votaries
and some hardy spirits experimented with some of the
green weed that decorated our waterline, dried in the
sun. But the most popular smoke was a combination of
ropeyarns, coffee-grounds and the bark off a pork-
barrel, rubbed up small and mixed in equal quantities.
If it had not the flavour of best Virginia it filled a place
and we were thankful.

We rolled on. The Trades, blowing light, set us
unduly far to the Westward and it was the 16th of the
month when we sighted the island of Fernando
Noronha. The island lies in about 4° south, not a
great distance off the Brazilian coast. It is most curiously shaped, very lofty, with symmetrical cone-shaped
mountains like volcanoes. The most conspicuous
feature is a bare conical hill about 1000ft. high in
the centre of the island. The place is used as a penal
settlement by the Brazilian Government, and passing
vessels are forbidden to call or make a landing on its
shores.

The Trades failed altogether soon after passing
Noronha and we ran into a region of light variable
winds. No rain fell, however, and we looked out

anxiously for the heavy downpours we might expect in these latitudes.

We drifted leisurely along, crossing the Line on the 19th and sighting a steamer bound out for the Plate the same afternoon. From her, after hoisting a signal and lowering a boat, we obtained a barrel of water, besides learning the news and acquiring an old newspaper or two.

We were ourselves the object of much interest on the part of the steamer's company. A wilder and more piratical boat's crew never boarded a peaceful merchant-man in mid-ocean. Lean, brown and sunburnt — bare-footed and half-naked to boot — with big Mac at the head, we followed the second mate aboard and swarmed over the bulwarks, eager for water, tobacco and news. Then we pulled back to the *Arethusa*, with a breaker of water and our odds and ends of treasures carefully stowed in the bottom of the boat.

The easy way we were making through the water gave us an opportunity to rig the grating the carpenter had made, and try to scrape some of the barnacles off our bottom. Lines were passed under the ship and bent on to each side of the grating, which was then hauled slowly back and fore under the hull, beginning for'ard and working aft. It was slow work, but successful in removing a great quantity of barnacles, even if the resultant increase in speed was scarcely appreciable.

Fishing was not neglected as we drifted peacefully along. We caught several fine albicore, a shark or two, and an absurd-looking creature called a sunfish. The latter was almost round, with two long fins and a stumpy tail. It was about twenty inches across, with a skin as prickly as a hedgehog's. As it lay on the deck, breathing heavily, it had a fussy important look, like a city alderman after a good dinner. We tried to dry and preserve it, but in a few days it made itself so objectionable that we returned it, considerably the worse for wear, to its native element.

We also captured several pretty little nautilus – Portugee men-o'-war, is their sea name. They are met with cruising about the equatorial regions in whole fleets. Nothing can be prettier than to see thousands upon thousands of these adventurous little mariners driving over summer seas before a gentle breeze. They look like a fleet of fairies – as though King Oberon had put to sea with all his court.

The light airs that flickered about the regions of the Doldrums just enabled the ship to ghost along, and forty-eight hours after crossing the Line, to our great relief, we had a heavy squall of rain. As soon as we saw it darkening the water to wind'ard all the scupper holes were blocked up, and with every tub and bucket we could lay hands on, we collected the water as it streamed off poop and foc'sle head. The first was brackish, through contact with the salt-stained decks and we put it into butts and breakers, but when it tasted quite fresh we emptied it down into our port tank, half filling the latter and removing all further fear of a shortage.

The rain only came just in time. Had it been deferred

a day or two longer we should have been in serious straits. As it was, we only had a depth of three inches of rusty and muddy water left in one tank. It was filthy, foul-smelling stuff and one had to be hard-pressed to drink it. Of all the hardships to be met with at sea, from thirst, good Lord! deliver me.

With the coming of the north-east Trades, a week after crossing the Line, we mended our pace a little and slanted steadily to the nor-nor-west. The North Star peeped over the horizon and cheered us with its homely twinkle that we had not seen for nearly a year. Every night the Southern Cross sank lower and lower astern and our conversation turned more and more towards the end of the voyage and what we should do when we got home.

The barque was now beginning to look shipshape and Bristol fashion and, by way of a finishing touch, the old man determined to paint the figurehead as it should be painted. I think I've mentioned that the *Arethusa* boasted an exceptionally fine one. It represented her namesake of old rising out of the waves, with long streaming hair and arm outstretched ahead. Hitherto this work of art had been hidden under a coat of dull mast-coloured paint. Now justice was to be done to it, and the saucy sea-maiden trimmed up with our best skill.

For this purpose we rigged a platform of stages under the bow and commenced operations. At about the twelfth attempt the bosun succeeded in mixing a flesh-colour paint to his satisfaction, and we gave her three coats of this. Then, with the expenditure of much pains, just the right tints were obtained for eyes, hair and lips and the nails of her fingers. The sea waves were painted a light-blue, edged with silver, and the scroll work above the cutwater crimson and gold. When it was done the saucy *Arethusa* looked her beauteous self and we were not a little proud of her. Indeed, when we got to port she came in for a good deal of admiration.

The next ten days were very uneventful ; all hands were hard at work painting, and the Trades did the rest. No sail broke the clear-cut circle of the horizon and we had the ocean to ourselves. Every night the North Star glittered higher above the foreyard, but we were ninety-three days out and it was high time we were getting on.

Crossing the Sargasso Sea we ran into huge fields of the strange weed that floats thereabouts. It drifts about in long wavy lines, or thick sluggish meadows, hither and thither before the wind.

The *Arethusa* ploughed her way through several fields many acres in extent. For two or three days, the surface of the sea, seen from aloft, was never free from long, fringing lines and isolated clumps of weed. Being first-voyagers, we, of course, fished for specimens of it and, dragging them on board, preserved pieces in bottles of salt-water to take home as curios. We caught great lumps by means of a grapnel of backstay wire, lashed on to a length of spun yarn. The weed is a yellowish-brown in colour, with large blob-like berries. These were yellow when we caught them, but at some seasons of the year assume a pink tinge and look very pretty.

When we reached the northern limits of the Sargasso Sea the weather broke up and became fresh and squally.

With the presage of rougher latitudes ahead we shifted sail. Heavy thunder squalls and vivid lightning were of constant occurrence. 'Royal drill' was resumed, and we boys got it a-plenty. To lay out on a royal yard with a thunder squall rattling about one's ears and the lightning ripping zig-zags all round, induces a tickling sensation in the bottom of one's spine. But ships' spars are seldom struck, and if ours were, we reflected, we shouldn't know much about it.

On the 24th of the month we sighted the Azores — the Western Isles in the language of the sea. In the neighbourhood of them, the Gulf Stream splits, part continuing to the north-east, and part turning to the south'ard to bend back into the North Equatorial current. Perhaps in consequence of this, the Isles seemed to be a dividing line between summer seas and our more boisterous northern weather. Bound south, one may hold wretched weather until the west point of Corvo is rounded; then, as if by magic, it clears up and the air becomes soft and summery.

It was Flores, the westernmost of the group, that on this occasion we sighted as we came sweeping up from mid-Atlantic. Fairest of all the group is Flores to an English seaman, for was it not off that island that Sir Richard Grenville bore down on the Spanish King's Armada of half a hundred sail and fought them till the

242

little *Revenge* was 'evened with the water'? No fault was it of his that the master gunner didn't split and sink the ship and forestall the glorious fate that came two days later and ended the gallantest fight in history. There never was a fight whose story stirs the blood like that of the *Revenge*, in all the long roll of England's sea-battles.

CHAPTER THIRTEEN

'The Fair Ports o' Home'

★

WITH the Western Isles astern began the last stage of our long homeward journey. We were one hundred and nine days out when we passed Flores – a very long passage, thanks to the battering we had received off the Horn, the light trades encountered and our barnacle encrusted hull.

The ship was as spick and span as paint and varnish could make her. Pots and brushes had been put away, and the hands were employed clearing out the forepeak and overhauling the bosun's stores. On the poop the old man still kept his fishing lines trailing astern and seemed loath to put the harpoons away. A school of porpoise that crossed our bow brought him forward with a run, and ere they passed he speared a big fellow from the knightheads. The eight-foot shaft went to splinters as the porpoise dived, but the line held and after a tremendous struggle we got it aboard.

The mate was not best pleased to have his white decks fouled with blood and slime. We hung the porpoise from our spare top-mast and cut steaks from it as an addition to our supper. As soon as it was suspended from the spar Nils ripped it open and cutting out its heart, which had hardly ceased to quiver, clapped it on the palm of his hand and swallowed it as though it had been an oyster.

This little touch was quite in keeping with the Finn's

general habits, for Nils was hardly civilized. He would eat almost anything. In Newcastle we got some barrels of stinking tallow with which to coat the ship's side, and I often observed Nils when he passed them, scoop out a lump with the end of his sheath knife and swallow it with intense relish. He would drink colza oil and paraffin if he got a chance and was indifferent whether the meat he ate was cooked or raw. What his religion was I never discovered, but he had never heard of the Bible. Always a bit of an enigma was Nils – heavily-built, smiling-faced, but rarely given to speaking. We called him a Russian Finn, but I think he was more probably a Lapp, for there was a Mongol-like appearance about his slanting eyes and high cheek bones.

All day and every day talk turned on when we should reach home, and guesses were made as to where we should go from Falmouth to discharge. Tommy, optimistic as ever, offered to bet anybody it would be London, but this we thought too good to be true and were inclined to think it would more likely be a port on the Continent.

We sighted several vessels as we drew near home waters, and were passed one day by a large Royal Mail liner, bound home from the West Indies. She swiftly overhauled us and we ran our ensign to the gaff-end as she forged up on the quarter, then hoisted our number, the code flags M K R W, that denoted our name. We saw her answering pennant fall swiftly to the dip, and thereafter signalled our port of departure, number of days out and the customary 'All's well,' knowing that in

a couple of days' time the information would appear
in the home papers and announce our impending
arrival to those who were scanning the shipping lists
for the *Arethusa's* name. The liner speedily forged
ahead, close enough for us to distinguish the passengers
lining her rail and to read the name *Trent* on her bows,
as she swept past.

By the 2nd of the month we estimated our position
to be, by dead reckoning, a few hundred miles from the
mouth of the Channel. We began to keep a sharp
look-out both for passing vessels and the first sight of
land. The breeze died down in the course of the day
and a film of mist spread over sea and sky. It gradually
grew thicker, till we were feeling our way across a grey
sea between eddying walls of fog, by the help of wandering airs that only just sufficed to give us steerage way.
The foghorn was sent for'ard and its mournful blast
quailed on the air every minute.

The air grew chilly and the damp, drifting mist that
covered everything was most depressing. Our progress,
moreover, was slow; we could hear nothing and see
nothing, and how long these conditions were going to
last there were no means of knowing.

Fog is common enough in the Channel and neigh-
bouring regions in the summer months and adds another
hazard to the navigation of those traffic-laden waters.
Far more dangerous to the mariner than gales is fog.
It is the most bewildering of the many difficulties with
which he has to contend. Passengers rarely seem to
realize this. An estimable old gentleman once asked

246

me, when the vessel we were in was joyously bucking into the north-east Trades blowing fresh, if the weather ever got any worse. The question took me so flat aback I hadn't wits enough to tell him it couldn't get any better. The same individual slept peacefully in his bunk all the way up Channel, when we ran into a thick fog and slowed down to about eight knots. He would, I've no doubt, have been surprised to learn that he was in a hundred times more danger than ever he was before.

As it was, the *Arethusa* had a narrow squeak and just escaped being sent to the bottom by a blundering German liner.

We were creeping blindly along, feeling our way down vistas of eddying fog-smoke, when we heard a pulsing throb and the low swish of water. We looked round and the little fog-horn on the foc'sle head coughed violently in response to Neilsen's hurried manipulating. Suddenly a giant shape loomed up out of the mist, standing directly down on top of us. Neilsen worked his wheezy toy of a fog-horn frantically and we jumped unbidden to the braces. There was no time to do anything, the steamer was almost into us and old Jamieson and the mate were heaving hard at the helm.

It was touch and go; only a few yards separated the two vessels, when there was a clatter of bells aboard the steamer and she swerved sharply to port. She missed our stern by only a few feet, travelling at near twenty knots. We read her name on the bows as she tore past – the *Deutschland* – at that time the crack vessel of the Hamburg-Amerika Line. For a moment she towered above us like a colossal wraith of the sea, then melted rapidly into the mist.

We came slowly back to course, cursing steamers in general, and the *Deutschland* in particular, with heart-felt vehemence.

This unlucky fog disgusted everybody. To be so near home and yet compelled to creep along at a snail's pace, unwitting where we were, was most tantalizing. Every watch we hoped to sight land, and watch after watch drew to an end and left us still disappointed. To add to our discomforts the cumbersome deep-sea-lead had to be hove every hour.

A slow, but very safe, method of navigation is this. The three L's — log, lead, and latitude — are the three points of the mariner's creed, and there is a cast-iron conviction among underwriters that with a proper use of the lead no vessel would ever be lost through stranding.

We used it *ad nauseam*; went through the performance every hour until we were sick of the sight of the long blue sinker and wet coils of line.

The mainyard would first be backed, stopping the way of the ship. The lead-line would then be passed along the weather side from aft forward, outside everything, and bent on to a twenty-eight pound sinker. The hands were stationed at intervals along the weather rail, each holding a few coils of slack in his hand.

At the order from aft, the man on the foc'sle head threw the lead overboard and ahead, crying out as he did so, 'Watch, there, watch!' As, each in turn, the men along the rail felt the weight of the lead, they let go the slack they were holding, repeating 'Watch, there, watch!' The old man stood on the weather quarter with the last of the line in his hand and took the sounding from there. It needed nice judgment to feel the moment the lead touched bottom and allow for the sag of the line.

As soon as the cast was taken the mainyard was squared again and the ship stood on her way. It was a wet, monotonous job and had to be carried out with wearisome regularity every time the bells marked the passage of the hours.

The results of the lead were very unsatisfactory ; we could not by their aid ascertain our position with any certainty; some of the men said we were heading for the Channel, others thought not. The skipper directed our course towards where he conceived the Scillies to be, and we crept along at the rate of a few knots. We were on soundings, as the hundred fathom line that fringes the British Isles and North-West Europe is called, and that is all we knew. The skipper, no doubt could have given a shrewd guess at our position, but everybody for'ard was quite in the dark and opinions varied from the neighbourhood of Ushant to St. George's Channel.

The appearance of the sea gave no indication. It might be thought that the shoaling of the water from the great depths of mid-ocean to a score or two of fathoms, would have been visible in its altered colour and appearance, but it is not so. The clammy fog made extended observation impossible and, as far as the surface was concerned, 'soundings' or ''mid-sea' looks much the same.

All through the 4th and 5th of May the numbing fog held. Under shortened sail, with a constant use of the lead, we drifted on, with eyes and ears alert, but still with no signs of land. We were almost beginning to think that the British Isles had sunk at their moorings and that nothing but water stretched endlessly in every direction. On the afternoon of the latter day the little breeze freshened, and by imperceptible degrees the fog began to melt away. Slowly the air grew clearer and clearer, bringing with it an ever-widening horizon. The sails now were kept steadily filled and the pleasant babble of water arose overside.

It was in the second dog-watch and just about dusk, as Beckett and I were sitting in the half-deck smoking our 'barrel blend,' that we heard a shout from for'ard and a stir among the men on deck. We made for the door and ran full tilt into Gilroy. His face was alight and his arm outstretched – 'There she is,' he said excitedly, 'England!'

We rushed out. There was still a pale clear light lingering over the sea's face to the ring of the horizon and as we ran to the rail to look ahead there winked, far away, a pin-point of light. Bright, lonely and unmistakable, it flashed for a moment and was gone. But it was enough. Yonder over the darkening sky-line was England and that light was the double flash of the Bishop, the farthest-flung outpost of her shores and the beacon-light of home!

My heart leapt within me to see this outermost Atlantic sentinel of my native land. All the werinesses of the last few days were forgotten and we gathered in little groups and fell a-talking of when we should reach Falmouth, where we should go to discharge and what we should do when we got home. The recurrent flash grew ever brighter and more distinct, full sail was piled on the old barque and course was altered to bring the Scillies abeam.

All that night we stood to the north-east, dipping along at not more than five knots, which was all the speed our weed-covered hull would let us attain to. Next morning we were abreast of the toe of England, with a number of fishing vessels in sight.

Off Mount's Bay we passed a warship, which spoke us and enquired our name and destination. She called us up by semaphore and put us in a fine pickle to answer her. Sailing ships do not — or rather did not — spend much time on signalling practice and were apt to be rusty in an emergency. They rarely had occasion for more than an infrequent flutter with the international code flags, and but for the fact that the captain had insisted on us practising 'flag wagging' on the passage home we should have been hard put to it to reply. As it was, Gilroy and I managed, with many requests to 'repeat,' to spell out her questions and made some sort of a shape — not too rapidly — at answering them.

However, the attempt was not bad for a windjammer and the warship was graciously pleased to accept our fair intentions for poor performance. Then we dipped our ensign to the little cruiser, the nautical courtesy equivalent to raising one's hat on shore, and stood on.

All that night under easy sail we cruised off and on. With the first streaks of dawn in the sky and the outline of the coast growing clearer to the nor'ard, we bore up and shaped a course for Falmouth. We picked up our pilot — a stalwart West-country-man — with round Saxon speech on his tongue — as we closed the land, feeling, as he clambered over the rail, that here at last was a tangible link with home and that our voyage was just about over and done with.

A fair and pleasant county is Cornwall, and justly beloved of sailors. The grand cliffs of the Lizard and Kynance, with the chequerboard of green fields about them, are most often the first speck of English land the returning wanderer sees. Enchanting as any fabled 'happy isle' or garden of the Hesperides are they to eyes long weary of wandering fields of barren foam. Yet no stretch of coast in all the world has been the scene of so many wrecks; it is a true saying, 'Paradise is under the shadow of swords.'

Past the Manacles we went, where the ill-fated *Mohegan* went ashore, to the little white town of Falmouth, lying snugly on the slopes of its beautiful bay.

Fair and green it looked on that bright spring morning. The harbour was full of shipping, both outward and homeward bound. A big Yankee barquentine was lifting her anchor as we came in, her white cotton canvas gleaming in the sun and the tinkle of the windlass pawls sounding musically on the air.

As soon as our anchor was down the captain went ashore and we were quickly boarded by a host of bum-boatmen with all sorts of things to sell. Good it was to taste tobacco again, and we sat round and smoked pipe pipe after pipe of strong black Irish roll in silent contentment.

Late in the afternoon the old man returned and order was given to unmoor. As we manned the windlass, word went round that our destination was Antwerp. The Belgian port was almost as handy for home as London, and, though Tommy had lost his bet, everyone was well satisfied. For the last time we piled the canvas on her and sheeted home with a will, slipping out as the shades of evening fell and the wink of the Eddystone guided us down the sea-way.

We had a spanking run up Channel, picking up our pilot, two days later, off Dungeness and making fast to a tug as it grew dusk. With the hawser aboard, one after another our kites were taken in and rolled up in a snug harbour-stow, while the Goodwins and the lights on the Kentish shore twinkled out to port. Flushing came next morning and then the monotonous passage of the Scheldt. We reached Antwerp at eleven o'clock that night and tied up with our nose on the mud in the Kempish dock, next to a beautiful barque, the *Indian Empire*, that had preceded us up the river, and had the satisfaction of hearing the mate's quiet: 'That'll do, you men' — our long voyage ended.

All hands were paid off next day and I bade a regretful farewell to Mr. Miller, the carpenter, Stedman, and

Mac. The two latter went off shoulder to shoulder, a crowd of small boys carrying their bags. The mate and the steward were staying by the ship and so, it seemed, to our loudly-voiced disappointment, were we apprentices. The following day, however, the captain told us we could go home and should be sent for when required. Our chests and bags were straightway packed and, piling them all into an open four-wheeler which almost broke down under the weight, we perched ourselves on top and drove off hilariously, catching the Harwich boat and arriving home in time for breakfast next morning.

Thanks to the captain's kindliness I had a full six weeks at home. After sixteen months of salt water one was able to appreciate such a holiday. And how glorious the English countryside looked that leafy month of June! Then one morning came a telegram, and I hurried off to Antwerp to rejoin the *Arethusa*.

Again we ringed the world and touched the ports desired. A very different voyage it was to the first — South Africa, India and the Far East — shipmates with death and divers disasters.

Six years in all, as apprentice and officer, I served in the old ship — the world mine oyster — before passing on to other rigs and finally settling staidly into steam.

And the gallant old barque herself? For a further half score years after I left her, under another commander and with other crews, she sailed the seas, ranging from the blue Caribbean to Cathay, and came at last to no inglorious end. Unlike so many of her sisters, it

'THE FAIR PORTS O' HOME'

was never her fate to spend a dishonoured old age under an alien flag, or to rust unwanted at her moorings in some remote backwater. Nor was she doomed to pass to Golden Harbour through storm at sea, succumbing at last to the elements she had so long mastered.

One summer's day in 1917, when homeward bound from the Mexican Gulf, in the grey of the morning off Eagle Island, she encountered a German submarine. It was the end of the old ship and she went down gloriously, like the daughter of the seas she was, with all sail set, the red ensign of England a-flutter at her peak and our old shark's tail pointing bravely to the stars.

PRINTED IN GUERNSEY, C.I., BRITISH ISLES,
BY THE STAR AND GAZETTE COMPANY LTD.

A LIST OF VOLUMES ISSUED IN THE TRAVELLERS' LIBRARY

3s. 6d. net each

JONATHAN CAPE LTD.

THIRTY BEDFORD SQUARE

LONDON

THE TRAVELLERS' LIBRARY

✱

A series of books in all branches of literature designed for the pocket, or for the small house where shelf space is scarce. Though the volumes measure only 7 inches by 4¾ inches, the page is arranged so that the margins are not unreasonably curtailed nor legibility sacrificed. The books are of a uniform thickness irrespective of the number of pages, and the paper, specially manufactured for the series, is remarkably opaque, even when it is thinnest.
A semi-flexible form of binding has been adopted, as a safeguard against the damage inevitably associated with hasty packing. The cloth is of a particularly attractive shade of blue and has the author's name stamped in gold on the back.

✱

1. CAN SUCH THINGS BE ?   A volume of Stories
    by Ambrose Bierce

¶ ' Bierce never wastes a word, never coins a too startling phrase ; he secures his final effect, a cold thrill of fear, by a simple, yet subtle, realism.   No anthology of short stories, limited to a score or so, would be complete without an example of his unique artistry.' *Morning Post*

2. THE BLACK DOG.   A volume of Stories
    by A. E. Coppard

¶ ' Mr. Coppard is a born story-teller.   The book is filled with a variety of delightful stuff : no one who is interested in good writing in general, and good short stories in particular, should miss it.' *Spectator*

3. THE AUTOBIOGRAPHY of a SUPER-TRAMP
    by W. H. Davies. With a preface by G. BERNARD SHAW

¶ Printed as it was written, it is worth reading for its literary style alone.   The author tells us with inimitable quiet modesty of how he begged and stole his way across America and through England and Wales until his travelling days were cut short by losing his right foot while attempting to ' jump ' a train.

4. BABBITT A Novel
by Sinclair Lewis

¶ 'One of the greatest novels I have read for a long time.'
H. G. Wells 　　　'*Babbitt* is a triumph.' *Hugh Walpole*
'His work has that something extra, over and above, which
makes the work of art, and it is signed in every line with the
unique personality of the author.' *Rebecca West*

5. THE CRAFT OF FICTION
by Percy Lubbock

¶ 'No more substantial or more charming volume of criticism
has been published in our time.' *Observer*
'To say that this is the best book on the subject is probably true;
but it is more to the point to say that it is the only one.'
Times Literary Supplement

6. EARLHAM
by Percy Lubbock

¶ 'The book seems too intimate to be reviewed. We want to be
allowed to read it, and to dream over it, and keep silence about
it. His judgment is perfect, his humour is true and ready; his
touch light and prim; his prose is exact and clean and full
of music.' *Times*

7. WIDE SEAS & MANY LANDS A Personal Narrative
by Arthur Mason.
With an Introduction by MAURICE BARING

¶ 'This is an extremely entertaining, and at the same time, moving
book. We are in the presence of a born writer. We read with
the same mixture of amazement and delight that fills us throughout a Conrad novel.' *New Statesman*

8. SELECTED PREJUDICES A book of Essays
by H. L. Mencken

¶ 'He is exactly the kind of man we are needing, an iconoclast,
a scoffer at ideals, a critic with whips and scorpions who does
not hesitate to deal with literary, social and political humbugs
in the one slashing fashion.' *English Review*

9. THE MIND IN THE MAKING   An Essay
     by James Harvey Robinson

¶ 'For me, I think James Harvey Robinson is going to be almost
as important as was Huxley in my adolescence, and William
James in later years.   It is a cardinal book.   I question whether
in the long run people may not come to it, as making a new
initiative into the world's thought and methods.' *From the
Introduction by* H. G. WELLS

10. THE WAY OF ALL FLESH   A Novel
     by Samuel Butler

¶ 'It drives one almost to despair of English Literature when one
sees so extraordinary a study of English life as Butler's posthumous
Way of All Flesh making so little impression.   Really, the
English do not deserve to have great men.' *George Bernard Shaw*

11. EREWHON   A Satire
     by Samuel Butler

¶ 'To lash the age, to ridicule vain pretension, to expose hypo-
crisy, to deride humbug in education, politics and religion, are
tasks beyond most men's powers; but occasionally, very
occasionally, a bit of genuine satire secures for itself more than a
passing nod of recognition. *Erewhon* is such a satire. . . . The
best of its kind since *Gulliver's Travels.' Augustine Birrell*

12. EREWHON REVISITED   A Satire
     by Samuel Butler

¶ 'He waged a sleepless war with the mental torpor of the pros-
perous, complacent England around him; a Swift with the
soul of music in him, and completely sane; a liberator of
humanity operating with the wit and malice and coolness of
Mephistopheles.' *Manchester Guardian*

13. ADAM AND EVE AND PINCH ME   Stories
     by A. E. Coppard

¶ Mr. Coppard's implicit theme is the closeness of the spiritual
world to the material; the strange, communicative sympathy
which strikes through two temperaments and suddenly makes
them one.   He deals with those sudden impulses under which
secrecy is broken down for a moment, and personality revealed
as under a flash of spiritual lightning.

14. DUBLINERS   A volume of Stories
by James Joyce

¶ A collection of fifteen short stories by the author of *Ulysses*. They are all of them brave, relentless, and sympathetic pictures of Dublin life; realistic, perhaps, but not crude; analytical, but not repugnant.   No modern writer has greater significance than Mr. Joyce, whose conception and practice of the short story is certainly unique and certainly vital.

15. DOG AND DUCK
by Arthur Machen

¶ 'As a literary artist, Mr. Arthur Machen has few living equals, and that is very far indeed from being his only, or even his greatest, claim on the suffrages of English readers.' *Sunday Times*

16. KAI LUNG'S GOLDEN HOURS
by Ernest Bramah

¶ 'It is worthy of its forerunner.   There is the same plan, exactitude, working-out and achievement; and therefore complete satisfaction in the reading.' *From the Preface by* HILAIRE BELLOC

17. ANGELS & MINISTERS, AND OTHER PLAYS
by Laurence Housman
Imaginary portraits of political characters done in dialogue—Queen Victoria, Disraeli, Gladstone, Parnell, Joseph Chamberlain, and Woodrow Wilson.

¶ 'It is all so good that one is tempted to congratulate Mr. Housman on a true masterpiece.' *Times*

18. THE WALLET OF KAI LUNG
by Ernest Bramah

¶ 'Something worth doing and done. . . . It was a thing intended, wrought out, completed and established.   Therefore it was destined to endure, and, what is more important, it was a success.' *Hilaire Belloc*

19. TWILIGHT IN ITALY
by D. H. Lawrence

¶ This volume of travel vignettes in North Italy was first published in 1916. Since then Mr. Lawrence has increased the number of his admirers year by year. In *Twilight in Italy* they will find all the freshness and vigour of outlook which they have come to expect from its author.

20. THE DREAM   A Novel
by H. G. Wells

¶ 'It is the richest, most generous and absorbing thing that Mr. Wells has given us for years and years.' *Daily News*
'I find this book as close to being magnificent as any book that I have ever read. It is full of inspiration and life.'
Daily Graphic

21. ROMAN PICTURES
by Percy Lubbock

¶ Pictures of life as it is lived—or has been or might be lived— among the pilgrims and colonists in Rome of more or less English speech.
'A book of whimsical originality and exquisite workmanship, and worthy of one of the best prose writers of our time.'
Sunday Times

22. CLORINDA WALKS IN HEAVEN
by A. E. Coppard

¶ 'Genius is a hard-ridden word, and has been put by critics at many puny ditches, but Mr. Coppard sets up a fence worthy of its mettle. He shows that in hands like his the English language is as alive as ever, and that there are still infinite possibilities in the short story.' *Outlook*

23. MARIUS THE EPICUREAN
by Walter Pater

¶ Walter Pater was at the same time a scholar of wide sympathies and a master of the English language. In this, his best known work, he describes with rare delicacy of feeling and insight the religious and philosophic tendencies of the Roman Empire at the time of Antoninus Pius as they affected the mind and life of the story's hero.

24. THE WHITE SHIP   Stories
    by Aino Kallas.
With an Introduction by JOHN GALSWORTHY
¶ 'The writer has an extraordinary sense of atmosphere.'
                    Times Literary Supplement
' Stories told convincingly and well, with a keen perception for
                natural beauty.' *Nation*

25. MULTITUDE AND SOLITUDE   A Novel
    by John Masefield
¶ ' As well conceived and done, as rich in observation of the
world, as profound where it needs to be profound, as any novel
                of recent writing.' *Outlook*
' This is no common book. It is a book which not merely
    touches vital things. It is vital.' *Daily News*

26. SPRING SOWING   Stories
    by Liam O'Flaherty
¶ 'Nothing seems to escape Mr. O'Flaherty's eye; his brain
turns all things to drama; and his vocabulary is like a river in
spate. *Spring Sowing* is a book to buy, or to borrow, or, yes,
                to steal.' *Bookman*

27. WILLIAM   A Novel
    by E. H. Young
¶ ' An extraordinary good book, penetrating and beautiful.'
                    Allan Monkhouse
' All its characters are very real and alive, and William himself
            is a masterpiece.' *May Sinclair*

28. THE COUNTRY OF THE POINTED FIRS
    by Sarah Orne Jewett
¶ ' The young student of American literature in the far distant
future will take up this book and say " a masterpiece ! " as
proudly as if he had made it.   It will be a message in a universal
language—the one message that even the scythe of Time spares.'
                From the Preface by WILLA CATHER
                                        * *

29. GRECIAN ITALY
by Henry James Forman

¶ 'It has been said that if you were shown Taormina in a vision you would not believe it. If the reader has been in Grecian Italy before he reads this book, the magic of its pages will revive old memories and induce a severe attack of nostalgia.' *From the Preface by* H. FESTING JONES

30. WUTHERING HEIGHTS
by Emily Brontë

¶ 'It is a very great book. You may read this grim story of lost and thwarted human creatures on a moor at any age and come under its sway.' *From the Introduction by* ROSE MACAULAY

31. ON A CHINESE SCREEN
by W. Somerset Maugham

¶ A collection of sketches of life in China. Mr. Somerset Maugham writes with equal certainty and vigour whether his characters are Chinese or European. There is a tenderness and humour about the whole book which makes the reader turn eagerly to the next page for more.

32. A FARMER'S LIFE
by George Bourne

¶ The life story of a tenant-farmer of fifty years ago in which the author of *The Bettesworth Book* and *The Memoirs of a Surrey Labourer* draws on his memory for a picture of the every-day life of his immediate forebears, the Smiths, farmers and handicraft men, who lived and died on the border of Surrey and Hampshire.

33. TWO PLAYS. *The Cherry Orchard & The Sea Gull*
by Anton Tchekoff. Translated by George Calderon

¶ Tchekoff had that fine comedic spirit which relishes the incongruity between the actual disorder of the world with the underlying order. He habitually mingled tragedy (which is life seen close at hand) with comedy (which is life seen at a distance). His plays are tragedies with the texture of comedy.

34. THE MONK AND THE HANGMAN'S DAUGHTER
by Adolphe Danziger de Castro and Ambrose Bierce

¶ 'They are stories which the discerning are certain to welcome. They are evidence of very unusual powers, and when once they have been read the reader will feel himself impelled to dig out more from the same pen.' *Westminster Gazette*

35. CAPTAIN MARGARET   A Novel
by John Masefield

¶ 'His style is crisp, curt and vigorous.   He has the Stevensonian sea-swagger, the Stevensonian sense of beauty and poetic spirit. Mr. Masefield's descriptions ring true and his characters carry conviction.' *The Observer*

36. BLUE WATER
by Arthur Sturges Hildebrand

¶ This book gives the real feeling of life on a small cruising yacht ; the nights on deck with the sails against the sky, long fights with head winds by mountainous coasts to safety in forlorn little island ports, and constant adventure free from care.

37. STORIES FROM DE MAUPASSANT
Translated by Elizabeth Martindale

¶ 'His "story" engrosses the non-critical, it holds the critical too at the first reading. . . .   That is the real test of art, and it is because of the inobtrusiveness of this workmanship, that for once the critic and the reader may join hands without awaiting the verdict of posterity.' *From the Introduction by* FORD MADOX FORD

38. WHILE THE BILLY BOILS   First Series
by Henry Lawson

¶ These stories are written by the O. Henry of Australia.   They tell of men and dogs, of cities and plains, of gullies and ridges, of sorrow and happiness, and of the fundamental goodness that is hidden in the most unpromising of human soil.

39. WHILE THE BILLY BOILS Second Series
by Henry Lawson

¶ Mr. Lawson has the uncanny knack of making the people he writes about almost violently alive. Whether he tells of jackeroos, bush children or drovers' wives, each one lingers in the memory long after we have closed the book.

41. IN MOROCCO
by Edith Wharton

¶ Morocco is a land of mists and mysteries, of trailing silver veils through which minarets, mighty towers, hot palm groves and Atlas snows peer and disappear at the will of the Atlantic cloud-drifts.

42. GLEANINGS IN BUDDHA-FIELDS
by Lafcadio Hearn

¶ A book which is readable from first page to last, and is full of suggestive thought, the essays on Japanese religious belief calling for special praise for the earnest spirit in which the subject is approached.

43. OUT OF THE EAST
by Lafcadio Hearn

¶ Mr. Hearn has written many books about Japan ; he is saturated with the essence of its beauty, and in this book the light and colour and movement of that land drips from his pen in every delicately conceived and finely written sentence.

44. KWAIDAN
by Lafcadio Hearn

¶ The marvellous tales which Mr. Hearn has told in this volume illustrate the wonder-living tendency of the Japanese. The stories are of goblins, fairies and sprites, with here and there an adventure into the field of unveiled supernaturalism.

45. THE CONQUERED
by Naomi Mitchison
A story of the Gauls under Cæsar

¶ 'With *The Conquered* Mrs. Mitchison establishes herself as the best, if not the only, English historical novelist now writing. It seems to me in many respects the most attractive and poignant historical novel I have ever read.' *New Statesman*

46. WHEN THE BOUGH BREAKS
by Naomi Mitchison
Stories of the time when Rome was crumbling to ruin

¶ 'Interesting, delightful, and fresh as morning dew. The connoisseur in short stories will turn to some pages in this volume again and again with renewed relish.' *Times Literary Supplement*

47. THE FLYING BO'SUN
by Arthur Mason
¶ 'What makes the book remarkable is the imaginative power which has re-created these events so vividly that even the supernatural ones come with the shock and the conviction with which actual supernatural events might come.' *From the Introduction by* EDWIN MUIR

48. LATER DAYS
by W. H. Davies
A pendant to *The Autobiography of a Super-Tramp*

¶ 'The self-portrait is given with disarming, mysterious, and baffling directness, and the writing has the same disarmingness and simpleness.' *Observer*

49. THE EYES OF THE PANTHER Stories
by Ambrose Bierce
¶ It is said that these tales were originally rejected by virtually every publisher in the country. Bierce was a strange man; in 1914 at the age of seventy-one he set out for Mexico and has never been heard of since. His stories are as strange as his life, but this volume shows him as a master of his art.

50. IN DEFENCE OF WOMEN
by H. L. Mencken
¶ 'All I design by the book is to set down in more or less plain form certain ideas that practically every civilized man and woman holds *in petto*, but that have been concealed hitherto by the vast mass of sentimentalities swathing the whole woman question.' *From the Author's Introduction*

51. VIENNESE MEDLEY   A Novel
by Edith O'Shaughnessy

¶ ' It is told with infinite tenderness, with many touches of grave or poignant humour, in a very beautiful book, which no lover of fiction should allow to pass unread.   A book which sets its writer definitely in the first rank of living English novelists.'
Sunday Times

52. PRECIOUS BANE   A Novel
by Mary Webb

¶ ' She has a style of exquisite beauty ; which yet has both force and restraint, simplicity and subtlety ; she has fancy and wit, delicious humour and pathos.   She sees and knows men aright as no other novelist does.   She has, in short, genius.'   *Mr. Edwin Pugh*

53. THE INFAMOUS JOHN FRIEND
by Mrs. R. S. Garnett

¶ This book, though in form an historical novel, claims to rank as a psychological study.   It is an attempt to depict a character which, though destitute of the common virtues of every-day life, is gifted with qualities that compel love and admiration.

54. HORSES AND MEN
by Sherwood Anderson

¶ '*Horses and Men* confirms our indebtedness to the publishers who are introducing his work here.   It has a unity beyond that of its constant Middle-west setting.   A man of poetic vision, with an intimate knowledge of particular conditions of life, here looks out upon a world that seems singularly material only because he unflinchingly accepts its actualities.'   *Morning Post*

55. SELECTED ESSAYS
by Samuel Butler

¶ This volume contains the following essays :

The Humour of Homer	How to Make the Best of Life
Quis Desiderio . . .?	The Sanctuary of Montrigone
Ramblings in Cheapside	A Medieval Girls' School
The Aunt, the Nieces, and the Dog	Art in the Valley of Saas
	Thought and Language

56. A POET'S PILGRIMAGE
by W. H. Davies

¶ *A Poet's Pilgrimage* recounts the author's impressions of his native Wales on his return after many years' absence. He tells of a walking tour during which he stayed in cheap rooms and ate in the small wayside inns. The result is a vivid picture of the Welsh people, the towns and countryside.

57. GLIMPSES OF UNFAMILIAR JAPAN. First Series
by Lafcadio Hearn

¶ Most books written about Japan have been superficial sketches of a passing traveller. Of the inner life of the Japanese we know practically nothing, their religion, superstitions, ways of thought. Lafcadio Hearn reveals something of the people and their customs as they are.

58. GLIMPSES OF UNFAMILIAR JAPAN. Second Series
by Lafcadio Hearn

¶ Sketches by an acute observer and a master of English prose, of a Nation in transition—of the lingering remains of Old Japan, to-day only a memory, of its gardens, its beliefs, customs, gods and devils, of its wonderful kindliness and charm—and of the New Japan, struggling against odds towards new ideals.

59. THE TRAVELS OF MARCO POLO
Edited by Manuel Komroff

¶ When Marco Polo arrived at the court of the Great Khan, Pekin had just been rebuilt. Kublai Khan was at the height of his glory. Polo rose rapidly in favour and became governor of an important district. In this way he gained first-hand knowledge of a great civilization and described it with astounding accuracy and detail.

60. SELECTED PREJUDICES. Second Series
by H. L. Mencken

¶ 'What a master of the straight left in appreciation! Everybody who wishes to see how common sense about books and authors can be made exhilarating should acquire this delightful book.'
Morning Post

61. THE WORLD'S BACK DOORS
by Max Murray
With an introduction by HECTOR BOLITHO

¶ This book is not an account so much of places as of people. The journey round the world was begun with about enough money to buy one meal, and continued for 66,000 miles. There are periods as a longshore man and as a sailor, and a Chinese guard and a night watchman, and as a hobo.

62. THE EVOLUTION OF AN INTELLECTUAL
by J. Middleton Murry

¶ These essays were written during and immediately after the Great War. The author says that they record the painful stages by which he passed from the so-called intellectual state to the state of being what he now considers to be a reasonable man.

63. THE RENAISSANCE
by Walter Pater

¶ This English classic contains studies of those 'supreme artists,' Michelangelo and Da Vinci, and of Botticelli, Della Robia, Mirandola, and others, who ' have a distinct faculty of their own by which they convey to us a peculiar quality of pleasure which we cannot get elsewhere.' There is no romance or subtlety in the work of these masters too fine for Pater to distinguish in superb English.

64. THE ADVENTURES OF A WANDERER
by Sydney Walter Powell

¶ Throwing up a position in the Civil Service in Natal because he preferred movement and freedom to monotony and security, the author started his wanderings by enlisting in an Indian Ambulance Corps in the South African War. Afterwards he wandered all over the world.

65. 'RACUNDRA'S' FIRST CRUISE
by Arthur Ransome

¶ This is the story of the building of an ideal yacht which would be a cruising boat that one man could manage if need be, but on which three people could live comfortably. The adventures of the cruise are skilfully and vividly told.

66. THE MARTYRDOM OF MAN
by Winwood Reade

¶ 'Few sketches of universal history by one single author have been written. One book that has influenced me very strongly is *The Martyrdom of Man*. This "dates," as people say nowadays, and it has a fine gloom of its own; but it is still an extraordinarily inspiring presentation of human history as one consistent process.' *H. G. Wells* in *The Outline of History*

67. THE AUTOBIOGRAPHY OF MARK RUTHERFORD
With an introduction by H. W. MASSINGHAM

¶ Because of its honesty, delicacy and simplicity of portraiture, this book has always had a curious grip upon the affections of its readers. An English Amiel, inheriting to his comfort an English Old Crome landscape, he freed and strengthened his own spirit as he will his reader's.

68. THE DELIVERANCE
by Mark Rutherford

¶ Once read, Hale White [Mark Rutherford] is never forgotten. But he is not yet approached through the highways of English letters. To the lover of his work, nothing can be more attractive than the pure and serene atmosphere of thought in which his art moves.

69. THE REVOLUTION IN TANNER'S LANE
by Mark Rutherford

¶ 'Since Bunyan, English Puritanism has produced one imaginative genius of the highest order. To my mind, our fiction contains no more perfectly drawn pictures of English life in its recurring emotional contrast of excitement and repose more valuable to the historian, or more stimulating to the imaginative reader.' *H. W. Massingham*

70. ASPECTS OF SCIENCE. First Series
   by J. W. N. Sullivan

¶ Although they deal with different aspects of various scientific
ideas, the papers which make up this volume do illustrate,
more or less, one point of view. This book tries to show one
or two of the many reasons why science may be interesting for
people who are not specialists as well as for those who are.

71. MASTRO-DON GESUALDO
   Giovanni Verga. Translated by D. H. Lawrence

¶ Verga, who died in 1922, is recognized as one of the greatest of
Italian writers of fiction. He can claim a place beside Hardy
and the Russians. 'It is a fine full tale, a fine, full picture of
life, with a bold beauty of its own which Mr. Lawrence must
   have relished greatly as he translated it.' *Observer*

72. THE MISSES MALLETT
   by E. H. Young

¶ The virtue of this quiet and accomplished piece of writing
lies in its quality and in its character-drawing; to summarize
it would be to give no idea of its charm. Neither realism nor
   romance, it is a book by a writer of insight and sensibility.

73. SELECTED ESSAYS. First Series
   by Sir Edmund Gosse, C.B.

¶ 'The prose of Sir Edmund Gosse is as rich in the colour of
young imagination as in the mellow harmony of judgment. Sir
Edmund Gosse's literary kit-kats will continue to be read with
avidity long after the greater part of the academic criticism of
the century is swept away upon the lumber-heap.' *Daily
   Telegraph*

74. WHERE THE BLUE BEGINS
   by Christopher Morley

¶ A delicious satirical fantasy, in which humanity wears a dog-
   collar.
'Mr. Morley is a master of consequent inconsequence. His
humour and irony are excellent, and his satire is only the more
salient for the delicate and ingenuous fantasy in which it is set.'
   Manchester Guardian

76. CONFESSIONS OF A YOUNG MAN
by George Moore

¶ 'Mr. Moore, true to his period and to his genius, stripped himself of everything that might stand between him and the achievement of his artistic object. He does not ask you to admire this George Moore. He merely asks you to observe him beyond good and evil as a constant plucked from the bewildering flow of eternity.' *Humbert Wolfe*

77. THE BAZAAR. Stories
by Martin Armstrong

¶ 'These stories have considerable range of subject, but in general they are stay-at-home tales, depicting cloistered lives and delicate finely fibred minds. . . . Mr. Armstrong writes beautifully.' *Nation and Athenæum*

78. SIDE SHOWS. Essays
by J. B. Atkins
With an Introduction by JAMES BONE

¶ Mr. J. B. Atkins was war correspondent in four wars, the London editor of a great English paper, then Paris correspondent of another, and latterly the editor of the *Spectator*. His subjects in *Side Shows* are briefly London and the sea.

79. SHORT TALKS WITH THE DEAD
by Hilaire Belloc

¶ In these essays Mr. Belloc attains his usual high level of pungent and witty writing. The subjects vary widely and include an imaginary talk with the spirits of Charles I, the barber of Louis XIV, and Napoleon, Venice, fakes, eclipses, Byron, and the famous dissertation on the Nordic Man.

80. ORIENT EXPRESS
by John dos Passos

¶ This book will be read because, as well as being the temperature chart of an unfortunate sufferer from the travelling disease, it deals with places shaken by the heavy footsteps of History, manifesting itself as usual by plague, famine, murder, sudden death and depreciated currency. Underneath the book is an ode to railroad travel.

81. SELECTED ESSAYS. Second Series
by Sir Edmund Gosse, C.B.

¶ A second volume of essays personally chosen by Sir Edmund Gosse from the wide field of his literary work. One is delighted with the width of his appreciation which enables him to write with equal charm on *Wycherley* and on *How to Read the Bible*.

82. ON THE EVE
by Ivan Turgenev. Translated by Constance Garnett

¶ In his characters is something of the width and depth which so astounds us in the creations of Shakespeare. *On the Eve* is a quiet work, yet over which the growing consciousness of coming events casts its heavy shadow. Turgenev, even as he sketched the ripening love of a young girl, has made us feel the dawning aspirations of a nation.

83. FATHERS AND CHILDREN
by Ivan Turgenev. Translated by Constance Garnett

¶ 'As a piece of art *Fathers and Children* is the most powerful of all Turgenev's works. The figure of Bazarov is not only the political centre of the book, but a figure in which the eternal tragedy of man's impotence and insignificance is realized in scenes of a most ironical human drama.' *Edward Garnett*

84. SMOKE
by Ivan Turgenev. Translated by Constance Garnett

¶ In this novel Turgenev sees and reflects, even in the shifting phases of political life, that which is universal in human nature. His work is compassionate, beautiful, unique; in the sight of his fellow-craftsmen always marvellous and often perfect.

85. PORGY. A Tale
by du Bose Heyward

¶ This fascinating book gives a vivid and intimate insight into the lives of a group of American negroes, from whom Porgy stands out, rich in humour and tragedy. The author's description of a hurricane is reminiscent in its power.

86. FRANCE AND THE FRENCH
by Sisley Huddleston

¶ 'There has been nothing of its kind published since the War. His book is a repository of facts marshalled with judgment; as such it should assist in clearing away a whole maze of misconceptions and prejudices, and serve as a sort of pocket encyclopædia of modern France.' *Times Literary Supplement*

88. CLOUD CUCKOO LAND. A Novel of Sparta
by Naomi Mitchison

¶ 'Rich and frank in passions, and rich, too, in the detail which helps to make feigned life seem real.' *Times Literary Supplement*

89. A PRIVATE IN THE GUARDS
by Stephen Graham

¶ In his own experiences as a soldier Stephen Graham has conserved the half-forgotten emotions of a nation in arms. Above all he makes us feel the stark brutality and horror of actual war, the valour which is more than valour, and the disciplined endurance which is human and therefore the more terrifying.

90. THUNDER ON THE LEFT
by Christopher Morley

¶ 'It is personal to every reader, it will become for every one a reflection of himself. I fancy that here, as always where work is fine and true, the author has created something not as he would but as he must, and is here an interpreter of a world more wonderful than he himself knows.' *Hugh Walpole*

91. THE MOON AND SIXPENCE
by Somerset Maugham

¶ A remarkable picture of a genius.
'Mr. Maugham has given us a ruthless and penetrating study in personality with a savage truthfulness of delineation and an icy contempt for the heroic and the sentimental.' *The Times*

92. THE CASUARINA TREE
by W. Somerset Maugham

❡ Intensely dramatic stories in which the stain of the East falls deeply on the lives of English men and women. Mr. Maugham remains cruelly aloof from his characters. On passion and its culminating tragedy he looks with unmoved detachment, ringing the changes without comment and yet with little cynicism.

93. A POOR MAN'S HOUSE
by Stephen Reynolds

❡ Vivid and intimate pictures of a Devonshire fisherman's life. ' Compact, harmonious, without a single—I won't say false—but uncertain note, true in aim, sentiment and expression, precise and imaginative, never precious, but containing here and there an absolutely priceless phrase. . . .' *Joseph Conrad*

94. WILLIAM BLAKE
by Arthur Symons

❡ When Blake spoke the first word of the nineteenth century there was none to hear it; and now that his message has penetrated the world, and is slowly re-making it, few are conscious of the man who first voiced it. This lack of knowledge is remedied in Mr. Symons' work.

95. A LITERARY PILGRIM IN ENGLAND
by Edward Thomas

❡ A book about the homes and resorts of English writers, from John Aubrey, Cowper, Gilbert White, Cobbett, Wordsworth, Burns, Borrow and Lamb, to Swinburne, Stevenson, Meredith, W. H. Hudson and H. Belloc. Each chapter is a miniature biography and at the same time a picture of the man and his work and environment.

96. NAPOLEON : THE LAST PHASE
by The Earl of Rosebery

❡ Of books and memoirs about Napoleon there is indeed no end, but of the veracious books such as this there are remarkably few. It aims to penetrate the deliberate darkness which surrounds the last act of the Napoleonic drama.

97. THE POCKET BOOK OF POEMS AND SONGS FOR THE OPEN AIR
Compiled by Edward Thomas

¶ This anthology is meant to please those lovers of poetry and the country who like a book that can always lighten some of their burdens or give wings to their delight, whether in the open air by day, or under the roof at evening ; in it is gathered much of the finest English poetry.

98. SAFETY PINS : ESSAYS
by Christopher Morley
With an Introduction by H. M. TOMLINSON

¶ Very many readers will be glad of the opportunity to meet Mr. Morley in the rôle of the gentle essayist. He is an author who is content to move among his fellows, to note, to reflect, and to write genially and urbanely ; to love words for their sound as well as for their value in expression of thought.

99. THE BLACK SOUL : A Novel
by Liam O'Flaherty

¶ '*The Black Soul* overwhelms one like a storm. . . . Nothing like it has been written by any Irish writer.' "*Æ*" in *The Irish Statesman*

100. CHRISTINA ALBERTA'S FATHER :
A Novel
by H. G. Wells

¶ ' At first reading the book is utterly beyond criticism ; all the characters are delightfully genuine.' *Spectator*
' Brimming over with Wellsian insight, humour and invention. No one but Mr. Wells could have written the whole book and given it such verve and sparkle.' *Westminster Gazette*

102. THE GRUB STREET NIGHTS ENTERTAINMENTS
by J. C. Squire

¶ Stories of literary life, told with a breath of fantasy and gaily ironic humour. Each character lives, and is the more lively for its touch of caricature. From *The Man Who Kept a Diary* to *The Man Who Wrote Free Verse*, these tales constitute Mr. Squire's most delightful ventures in fiction ; and the conception of the book itself is unique.

103. ORIENTAL ENCOUNTERS
by Marmaduke Pickthall

❡ In *Oriental Encounters*, Mr. Pickthall relives his earlier manhood's discovery of Arabia and sympathetic encounters with the Eastern mind. He is one of the few travellers who really bridges the racial gulf.

105. THE MOTHER: A Novel
by Grazia Deledda
With an introduction by D. H. LAWRENCE

❡ An unusual book, both in its story and its setting in a remote Sardinian hill village, half civilized and superstitious. The action of the story takes place so rapidly and the actual drama is so interwoven with the mental conflict, and all so forced by circumstances, that it is almost Greek in its simple and inevitable tragedy.

106. TRAVELLER'S JOY: An Anthology
by W. G. Waters

❡ This anthology has been selected for publication in the Travellers' Library from among the many collections of verse because of its suitability for the traveller, particularly the summer and autumn traveller, who would like to carry with him some store of literary provender.

107. SHIPMATES: Essays
by Felix Riesenberg

❡ A collection of intimate character portraits of men with whom the author has sailed on many voyages. The sequence of studies blends into a fascinating panorama of living characters.

108. THE CRICKET MATCH
by Hugh de Selincourt

❡ Through the medium of a cricket match the author endeavours to give a glimpse of life in a Sussex village. First we have a bird's-eye view at dawn of the village nestling under the Downs; then we see the players awaken in all the widely different circumstance of their various lives, pass the morning, assemble on the field, play their game, united for a few hours, as men should be, by a common purpose—and at night disperse.

109. RARE ADVENTURES AND PAINEFULL
PEREGRINATIONS (1582–1645)
by William Lithgow
Edited, and with an Introduction by B. I. LAWRENCE

¶ This is the book of a seventeenth-century Scotchman who walked over the Levant, North Africa and most of Europe, including Spain, where he was tortured by the Inquisition. An unscrupulous man, full of curiosity, his comments are diverting and penetrating, his adventures remarkable.

110. THE END OF A CHAPTER
by Shane Leslie

¶ In this, his most famous book, Mr. Shane Leslie has preserved for future generations the essence of the pre-war epoch, its institutions and individuals. He writes of Eton, of the Empire, of Post-Victorianism, of the Politicians. . . . And whatever he touches upon, he brilliantly interprets.

111. SAILING ACROSS EUROPE
by Negley Farson
With an Introduction by FRANK MORLEY

¶ A voyage of six months in a ship, its one and only cabin measuring 8 feet by 6 feet, up the Rhine, down the Danube, passing from one to the other by the half-forgotten Ludwig's Canal. To think of and plan such a journey was a fine imaginative effort and to write about it interestingly is no mean accomplishment.

112. MEN, BOOKS AND BIRDS—Letters to a friend
by W. H. Hudson
With Notes, some Letters, and an Introduction by
MORLEY ROBERTS

¶ An important collection of letters from the naturalist to his friend, literary executor and fellow-author, Morley Roberts, covering a period of twenty-five years.

113. PLAYS ACTING AND MUSIC
by Arthur Symons

¶ This book deals mainly with music and with the various arts of the stage. Mr. Arthur Symons shows how each art has its own laws, its own limits; these it is the business of the critic jealously to distinguish. Yet in the study of art as art, it should be his endeavour to master the universal science of beauty.

114. ITALIAN BACKGROUNDS
by Edith Wharton

¶ Mrs. Wharton's perception of beauty and her grace of writing are matters of general acceptance. Her book gives us pictures of mountains and rivers, monks, nuns and saints.

115. FLOWERS AND ELEPHANTS
by Constance Sitwell. With an Introduction by E. M. Forster

¶ Mrs. Sitwell has known India well, and has filled her pages with many vivid little pictures, and with sounds and scents. But it is the thread on which her impressions are strung that is so fascinating, a thread so delicate and rare that the slightest clumsiness in definition would snap it.

116. THE MOON OF THE CARIBBEES : and Other Plays of the Sea
by Eugene O'Neill. With an Introduction by St. John Ervine

¶ 'Mr. O'Neill is immeasurably the most interesting man of letters that America has produced since the death of Walt Whitman.' *From the Introduction.*

117. BETWEEN EARTH AND SKY. Stories of Gypsies
by Konrad Bercovici. With an Introduction by A. E. Coppard

¶ Konrad Bercovici, through his own association with gipsies, together with a magical intuiton of their lives, is able to give us some unforgettable pictures of those wanderers who, having no home anywhere, are at home everywhere.

118. THE HOUSE WITH THE GREEN SHUTTERS
by George Douglas. With an Introduction by J. B. Priestley

¶ This powerful and moving story of life in a small Scots burgh is one of the grimmest studies of realism in all modern fiction. The author flashes a cold and remorseless searchlight upon the backbitings, jealousies, and intrigues of the townsfolk, and his story stands as a classic antidote to the sentimentalism of the kailyard school.

119. FRIDAY NIGHTS
by Edward Garnett

¶ Of *Friday Nights* a *Times* reviewer wrote : ' Mr. Garnett is " the critic as artist," sensitive alike to elemental nature and the subtlest human variations. His book sketches for us the possible outlines of a new humanism, a fresh valuation of both life and art.'

120. DIVERSIONS IN SICILY
by Henry Festing Jones

¶ Shortly before his sudden and unexpected death, Mr. Festing Jones chose out *Diversions in Sicily* for reprinting in the Travellers' Library from among his three books of mainly Sicilian sketches and studies. The publishers hope that the book, in this popular form, will make many new friends. These chapters, as well as any that he wrote, recapture the wisdom, charm, and humour of their author.

121. DAYS IN THE SUN: A Cricketer's Book.
by Neville Cardus ('Cricketer' of the *Manchester Guardian*).

¶ These sketches were first published in *A Cricketer's Book* (1922) and in *Days in the Sun* (1924), they have now been revised for re-issue in *The Travellers' Library*. The author says 'the intention of this book is modest – it should be taken as a rather freely compiled journal of happy experiences which have come my way on our cricket fields.'

122. COMBED OUT
by F. A. Voigt

¶ This account of life in the army in 1917–18 both at home and in France is written with a telling incisiveness. The author does not indulge in an unnecessary word, but packs in just the right details with an intensity of feeling that is infectious.

123. CONTEMPORARIES OF MARCO POLO
edited by Manuel Komroff

¶ This volume comprises the Travel Records in the Eastern parts of the world of William of Rubruck (1253–1255), the Journey of John of Pian de Carpini (1245–1247), the Journey of Friar Odoric (1318–1330), the Oriental Travels of Rabbi Benjamin of Tudela (1160–1173). They describe the marvels and wonders of Asia under the Khans.

124. TENNYSON
by Hugh I'Anson Fausset

¶ Mr. Fausset speaks of Tennyson on his deathbed as 'the monument of the conscience and the poetry of more than half a century,' and his study of his qualities as poet, man, and moralist is by implication a study of some of the predominant characteristics of the Victorian age. His book, however, is as pictorial as it is critical, being woven, to quote *The Times*, 'like an arras of delicate colour and imagery.' It has been revised for issue in 'the Travellers' Library' and a new preface added.

125. CAPTIVES OF TIPU: SURVIVORS' NARRATIVES
edited by A. W. Lawrence

¶ Three records of heroic endurance, which were hitherto unobtainable at a reasonable price. In addition to the well-known stories of Bristow and Scurry, a soldier and a seaman, who were forcibly Mohammedanized and retained in the service of Mysore till their escape after ten years, extracts are given from an officer's diary of his close imprisonment at Seringapatam.

126. MEMOIRS OF A SLAVE-TRADER
by Theodore Canot. Set down by Brantz Mayer and now edited by A. W. Lawrence

¶ In 1854 a cosmopolitan adventurer, who knew Africa at the worst period of its history, dictated this sardonic account of piracy and mutiny, of battles with warships or rival traders, and of the fantastic lives of European and half-caste slavers on the West Coast.

127. BLACK LAUGHTER
by Llewelyn Powys. Author of *Ebony and Ivory*, etc.

¶ *Black Laughter* is a kind of *Robinson Crusoe* of the continent of Africa. Indeed, Llewelyn Powys resembles Daniel Defoe in the startlingly realistic manner in which he conveys the actual feelings of the wild places he describes. You actually share the sensations of a sensitive and artistic nature suddenly transplanted from a peaceful English village into the heart of Africa.

128. THE INFORMER
by Liam O'Flaherty. Author of *Spring Sowing*, etc.

¶ This realistic novel of the Dublin underworld is generally conceded to be Mr. O'Flaherty's most outstanding book. It is to be produced as a film by British International Pictures, who regard it as one of the most ambitious of their efforts.

129. THE BEADLE. A novel of South Africa
by Pauline Smith. Author of *The Little Karoo*

¶ 'A story of great beauty, and told with simplicity and tenderness that makes it linger in the memory. It is a notable contribution to the literature of the day.' *Morning Post.*

130. FISHMONGER'S FIDDLE. Short Stories
by A. E. Coppard. Author of *The Black Dog*, *Silver Circus*, etc.

¶ 'In definite colour and solid strength his work suggests that of the old Dutch Masters. Mr. Coppard is a born story-teller.' *Times Literary Supplement.*

★

Note

The *Travellers' Library* is now published as a joint enterprise by Jonathan Cape Ltd. and William Heinemann Ltd. The new volumes announced here to appear during the summer of 1929 include those to be published by both firms. The series as a whole or any title in the series can be ordered through booksellers from either Jonathan Cape or William Heinemann. Booksellers' only care must be not to duplicate their orders.

Made and Printed in Great Britain by Butler & Tanner Ltd., Frome and London

Date Due

Printed in the USA
CPSIA information can be obtained
at www.ICGtesting.com
CBHW071052140424
6901CB00014B/934

9 781014 790088

EX LIB-RIS

EATON HALL